101
DIFFERENT WAYS TO
BUILD HOMES
AND PENS
FOR YOUR ANIMALS

A COMPLETE STEP-BY-STEP GUIDE

By Randy LaTour, Industrial Designer,
and Sarah Ann Beckman, Co-Author

101 DIFFERENT WAYS TO BUILD HOMES AND PENS FOR YOUR ANIMALS: A COMPLETE STEP-BY-STEP GUIDE

Copyright © 2011 Atlantic Publishing Group, Inc.
1405 SW 6th Avenue • Ocala, Florida 34471 • Phone 800-814-1132 • Fax 352-622-1875
Web site: www.atlantic-pub.com • E-mail: sales@atlantic-pub.com
SAN Number: 268-1250

LaTour, Randy, 1959-
  101 different ways to build homes and pens for your animals : a complete step-by-step guide / by Randy LaTour.
    p. cm.
  Includes bibliographical references and index.
  ISBN-13: 978-1-60138-371-6 (alk. paper)
  ISBN-10: 1-60138-371-1 (alk. paper)
  1. Animal housing. I. Title. II. Title: One hundred one different ways to build homes and pens for your animals.
  SF91.L38 2010
  636.08'31--dc22

                              2010045856

PROJECT MANAGER: Amy Moczynski • amoczynski@atlantic-pub.com
INTERIOR LAYOUT: Antoinette D'Amore • addesign@videotron.ca
PROOFREADER: Crystal McKenna • crystalmckenna@me.com
COVER DESIGN: Meg Buchner • meg@megbuchner.com
BACK COVER DESIGN: Jackie Miller • millerjackiej@gmail.com

Printed in the United States

Printed on Recycled Paper

We recently lost our beloved pet "Bear," who was not only our best and dearest friend but also the "Vice President of Sunshine" here at Atlantic Publishing. He did not receive a salary but worked tirelessly 24 hours a day to please his parents. Bear was a rescue dog that turned around and showered myself, my wife, Sherri, his grandparents Jean, Bob, and Nancy, and every person and animal he met (maybe not rabbits) with friendship and love. He made a lot of people smile every day.

We wanted you to know that a portion of the profits of this book will be donated to The Humane Society of the United States. —*Douglas & Sherri Brown*

The human-animal bond is as old as human history. We cherish our animal companions for their unconditional affection and acceptance. We feel a thrill when we glimpse wild creatures in their natural habitat or in our own backyard.

Unfortunately, the human-animal bond has at times been weakened. Humans have exploited some animal species to the point of extinction.

The Humane Society of the United States makes a difference in the lives of animals here at home and worldwide. The HSUS is dedicated to creating a world where our relationship with animals is guided by compassion. We seek a truly humane society in which animals are respected for their intrinsic value, and where the human-animal bond is strong.

Want to help animals? We have plenty of suggestions. Adopt a pet from a local shelter, join the HSUS, and be a part of our work to help companion animals and wildlife. You will be funding our educational, legislative, investigative, and outreach projects in the U.S. and across the globe.

Or perhaps you'd like to make a memorial donation in honor of a pet, friend or relative? You can through our Kindred Spirits program. And if you'd like to contribute in a more structured way, our Planned Giving Office has suggestions about estate planning, annuities, and even gifts of stock that avoid capital gains taxes.

Maybe you have land that you would like to preserve as a lasting habitat for wildlife. Our Wildlife Land Trust can help you. Perhaps the land you want to share is a backyard—that's enough. Our Urban Wildlife Sanctuary Program will show you how to create a habitat for your wild neighbors.

So you see, it's easy to help animals. And The HSUS is here to help.

THE HUMANE SOCIETY
OF THE UNITED STATES.

2100 L Street NW • Washington, DC 20037 • 202-452-1100
www.hsus.org

Dedication

I would like to dedicate this book to my father, who inspired me to be creative and who taught me how to make something from nothing. I remember growing up as a little boy and watching my father work on handyman projects around the house. He would take scrap material and transform it into a useful item. If anything was broken, he could fix it. While others may discard a broken chair, he would find a way to salvage and repair the chair, adding 20 years to its life. Thanks for the life lessons.

Author biography: Randy LaTour began working as a design engineer immediately following college. With more than 20 years of design experience, he has created everything from custom machinery with complex material handling systems to simple structures such as tree house plans. One of his more notable design accomplishments is engineering, fabricating, and piloting a championship-winning racecar.

Introduction

This book was created for pet owners and animal lovers as a construction guide for building your animal friend a comfortable home or habitat. The following chapters will provide step-by-step instructions that will allow the do-it-yourself handyperson to complete the project on his or her own. No matter what kind of pet you own, you will be able to choose your favorite plan to house your animal companion from one of this book's 101 unique designs.

There are many benefits to building your pet's home on your own, the first of which is the cost savings. If you follow this book's detailed plans and perform the labor yourself, your reward will be substantial cost savings. You will discover how to choose the best plan for your animal and learn construction and safety tips. This book will offer you design plans for homes for dogs, cats, birds, fish, turtles, snakes, rabbits, chickens, and more. In addition to the financial savings, taking on a project of this magnitude will be a fun and rewarding experience for the whole family. Get the children involved in most aspects of construction, from selecting a plan and choosing a color, to building the animal home and driving in the last nail. Cap off your project with a nameplate, toys, and special accessories for your favorite pet.

Upon successfully completing this family project, parents, children, and pets will become closer, and your pet will enjoy its new home for years to come.

BEFORE YOU GET STARTED

As you prepare for construction of your new animal home, there are a few things to consider. The No. 1 consideration is safety. If you have experience working with tools and building, you may already be familiar with basic safety procedures. However, while most everyone has used a hammer at some point in his or her life, this may be your first construction project or your first time using a power tool. If this is your first woodworking venture, be sure to read the section on power tool safety for an overview of how to safely and properly use tools and equipment.

The second thing you will need to consider is selection og the proper home for your animal. Each chapter will provide several design options for each animal for you to review. Choose the best one for your climate, the size of your pet, and the style that matches your pet's environment or the architectural style of your home.

Finally, after you have selected a plan, you may need to obtain building permits and/or ensure you abide by applicable building codes. In many rural areas, you may be allowed to build without applying for a permit. However, in some residential areas, you may need a building permit for something as small as a doghouse. Some community covenants may even restrict the size or color of any outdoor structure in the neighborhood. While dogs, cats, and birds are commonplace within the city limits, your neighborhood or city may have restrictions on chickens, reptiles, and other undomesticated animals.

To be on the safe side, it is imperative to research your local area for potential restrictions. Start with a visit to your local town hall or building department. You may even be able to find a website for your city or county that accepts online applications. You can perform an online search for your local department of building inspection or the department that handles building permits to see if there is any information on the Internet. The local building inspector can provide guidance as to whether or not your project will require permits, or if it is even allowed at all. The office of the building inspector is most often located at your city or town hall.

Tools and Equipment

When building a new home for your domestic pet or farm animal, you will use basic tools throughout the construction process. Depending on the scope or magnitude of your project, this may entail using basic hand tools, power tools, or specialty equipment. This chapter provides a fundamental overview of the tools you may need for your project.

Claw hammer: A 13-ounce claw hammer is the best choice for assembling the 2x4 framework for your project. A hammer that is too small will make it difficult to drive in the structural nails used for framework, such as 10d 3-inch nails. A hammer that is too heavy will cause fatigue, and you will never make it through the day.

Roofing hammer: While a claw hammer is acceptable for installing roofing shingles, a standard roofing hammer will make the job much easier. A roofing hammer is also known as a roofing hatchet. The magnetic end of the roofing hatchet will hold the nail in place for a one-hit installation, greatly decreasing the time it takes to complete a project. The hatchet end can be used for splitting wood shingles as they are installed.

Framing level: To ensure that your finished project is pleasing to the eye, make sure each component is level before and after installation. A 4-foot level is the best tool for leveling the main structure, as it will provide a more accurate reading on longer lengths of lumber, while a 12-inch torpedo level or a 24-inch framing level can be used for installing smaller components such as brackets and doggie doors.

Framing square: Final assembly and fitting matching components together will be much smoother and less aggravating if you cut each individual component square and true. Use a typical 24-inch steel framing square and draw a cut line on the wood using a soft pencil or a metal scribe. A metal scribe looks like a small screwdriver with the blade ground down to form a sharp point.

Screwdriver: Wood screws typically require a No. 2 or No. 3 Phillips-head screwdriver for installation. To expedite your project, use a Phillips-head adapter in your cordless drill.

Hand drill: When drilling pilot holes in the wood prior to installing wood screws, use a 3/8-inch hand drill. A pilot hole will prevent the wood from splitting as you insert the screws. To make a pilot hole, pre-drill a hole that is 1/16 inch to 1/8 inch smaller than the screw size being used. Using a rechargeable cordless drill will be more convenient. Today's battery-operated cordless drills provide the same power as conventional electric hand drills without the hassle of using an extension cord.

Tape measure: Use a steel tape measure to properly size each component. Double-check each measurement prior to making your final cut. A 16-foot tape measure will be adequate for most projects in this book. A ¾-inch-wide tape measure is fairly rigid and good for a long reach without the tape collapsing under its own weight.

Circular saw: A standard powered circular saw with a typical 7¼-inch blade will be used to cut framing to length or to cut plywood to size. Use a fine tooth blade for plywood, siding, and cross cuts and a coarse blade for ripping long lengths of dimensional lumber.

Sawhorse: A sawhorse is simply a support stand consisting of a horizontal 2x4 about 3 feet long supported by four 2x4 legs. Using a sawhorse is like having a helper to support the wood as you make your final cut. Use two, three, or even four sawhorses when cutting plywood or framing lumber. When cutting a large 4-foot by 8-foot sheet of plywood, it is best to use four sawhorses to support both components.

Clamps: Bar clamps and C-clamps will act like an extra pair of hands when assembling components together.

Safety Considerations

Working with power tools and even hand tools can create hazards for the handyperson and his or her crew. Even a simple stepladder can produce a level of danger while working at an elevated height. Follow these safety tips and use proper safety equipment to ensure your project is safe and uneventful. While many of us have received some form of on-the-job safety training at our workplace, we rarely practice what we have learned when performing projects at home. Following a few simple safety rules can prevent a serious injury and provide greater enjoyment as you work on your project.

Ladder safety: Always choose the correct ladder for the task at hand. If using a stepladder, never stand on the top two rungs of the ladder. The ladder can become unstable, and you may lose your balance.

An extension ladder must be tall enough to reach the work area and should extend 3 feet above the base of the roof if you will be accessing the roof. By extending the ladder above the roof, you will have a place to hold on and maintain your balance as you transition from the ladder rungs to the roof surface. Extension ladders should be tied off at the top, which means you must use a rope or strap to secure the top of the ladder to keep it from sliding from side to side. The proper angle for an extension ladder is 1 foot from the wall for every 3 to 4 feet of elevation.

Never lean out from side to side on a ladder; always keep your navel between the rungs. As your hips extend beyond the side of the ladder, the ladder itself may become unstable. Never carry tools or equipment up a ladder; use a rope to hoist materials to the roof.

Never allow more than one person on a ladder at the same time, and always face the ladder when ascending or descending. Always keep a three-point contact as you climb the ladder: either one foot and two hands or two feet and one hand should be in contact with the ladder at all times.

Power Tool Safety and Personal Protective Equipment (PPE)

When using power tools, always wear appropriate PPE. If you are using a circular saw, this will include safety glasses or goggles, the sunglasses you wear to the beach will not suffice. Be sure to wear proper glasses with an approved ANSI Z81 rating stamped on the frame or lens. The ANSI rating is established by the American National Standards Institute to ensure your protective gear is suitable to use in potentially hazardous conditions.

Gloves: Wear appropriate leather gloves when handling sharp objects or wood to avoid cuts and splinters. Select a pair with a double-layered palm for extra protection. Be sure to shop around as the price of gloves can range from a few dollars per pair to as much as $20 per pair for mechanics' gloves. Wear gloves when handling pressure-treated wood to avoid contact with harmful chemicals. If you do come in contact with pressure-treated wood, wash your hands thoroughly prior to consuming food.

Head protection: When working on large projects with elevated overhead work, a proper hard hat should be worn at all times. Those working on the ground should keep clear of overhead activity.

Dust masks: Wear a dust mask to reduce the risk of inhaling small wood particles when cutting wood.

Proper lifting: When handling large sheets of plywood or long lengths of dimensional lumber, work in pairs. A two-person team will make lifting much easier. To avoid back injury, lift with your knees and avoid twisting your body.

Extension cords: When working outside, be sure to connect your power cord to a ground fault circuit interrupter (GFCI) outlet. The GFCI will protect you from shock. Most homes have a GFCI circuit on the exterior of the home, as it is required by code. If you have an older home or farmhouse without a GFCI outlet, you can purchase an extension cord with a built-in GFCI and plug it into any standard household outlet. Use a heavy extension cord with a minimum of 14-gauge wire.

Materials and Hardware

Selecting the proper materials for your projects is essential for a long-lasting project. While pressure-treated lumber will provide longevity for components that contact the ground, avoid using it for components that may come in contact with your pet. While pressure-treated lumber is much safer to work with today than it was in the past, the chemicals used to treat the wood still may be harmful if consumed.

Framing material: The typical construction material for projects is standard 2x4 or 2x6 pine. Pine is one of the most common and cost-effective types of wood available for construction. While the nominal dimensions are 2x4, the actual size of a 2x4 is 1½ inches by 3½ inches. This information is critical when calculating dimensions, and something that is often overlooked when planning projects. See the following chart for a list of standard sizes.

Dimensional Lumber	
Thickness by Width (inches)	
Nominal Size	Actual Size
2x2	1½" x 1½"
2x4	1½" x 3½"
2x6	1½" x 5½"
2x8	1½" x 7¼"
2x10	1½" x 9¼"
1x2	¾" x 1½"
1x4	¾" x 3½"
1x6	¾" x 5½"
1x8	¾" x 7¼"
1x10	¾" x 9¼"
Note: Pressure-treated wood can be up ¼" thicker.	

Paint: Use a high-quality exterior paint and primer when working on the projects in this book. Also, make sure you do not use lead-based paint. While lead-based paint is no longer sold in North America, it may be found in other parts of the world. Clean loose dust and debris from wood surfaces prior to painting. You should allow pressure-treated lumber to age for 90 days to six months prior to painting to allow the chemicals on the surface of the lumber to escape. If you paint pressure-treated lumber too soon, the paint will not adhere to the surface.

Hardware: To ensure your projects stand up over time, assemble your projects using screws instead of nails. When assembling pressure-treated components, be sure to purchase coated screws suitable for pressure-treated use. For non-pressure-treated wood, galvanized screws will suffice.

Hinges and latches: Purchase galvanized hinges and latches suitable for outdoor use. The galvanized coating on steel hinges and hardware will prevent the components from rusting when exposed to outdoor weather conditions. Powder-coated hardware, or hardware coated with a durable paint finish that is baked on at the factory, is a suitable alternative.

Permits and Building Codes

For the majority of the projects in this book, you can proceed without needing a building permit. However, in some residential areas and particularly for larger projects, you will need to check your local building codes and review your neighborhood covenants to see if your project is allowed or if permits may be required.

In the case of a large structure, you may be required to submit your drawings to the local code office before construction begins. You also may need to obtain an approval stamp from a local architect for concrete work and framing work. The certified architect can determine the proper size of components and correct concrete thickness applicable to your region. To avoid frustration, it is best to consult your local building department or code enforcement office before you begin construction.

Now that you know a bit about the tools and equipment you will need to get started, you will learn how to make a few projects for man's best friend.

BUILDING A DOGHOUSE

As "man's best friend," dogs are high on the list of popular pets. A dog quickly becomes part of any family thanks to its loyalty, fun-loving nature, instincts to protect, and an uncanny intuition as to what humans are feeling and need. With hundreds of domesticated dog breeds in various sizes, temperaments, furriness, and playfulness, each person has his or her own opinion about the best dog breed. While some people love a dog that is content sitting on their lap all day, others want an active canine that loves running in the park and playing catch.

The dog's strong tie with humans is the result of a relationship that started around 15,000 years ago.

Throughout history, dogs provided humans with various practical purposes, including hunting, guarding, mountain and water rescue, farming, and herding. For some cultures, such as the Native American culture, dogs played a religious role, helping deceased humans navigate the afterlife; it was the belief in many cultures that dogs served as "escorts" in death. While dogs still fulfill practical purposes today, their place with humans has shifted to emphasize companionship. Many people consider a dog as a best friend or like a child. Some dogs visit the hair salon frequently, wear clothes, and find themselves featured on family holiday cards.

A dog considers itself to be a part of a family because of its natural instincts as a pack animal. Scientists believe all dogs descend from the wolf, *Canis lupus*. Wolves are well-known for traveling in groups with a set hierarchy so much so that humans often call a threatening group of people a "pack of wolves." Dogs, by instinct, are territorial animals. This derives from their ties with wolves, which are extremely territorial. To a domesticated dog, your family is its pack, and your house is its territory. As your dog's pack leader, it is your responsibility to make sure your dog is well cared for.

Despite sharing so many characteristics with the wild wolf, the 15,000 years that domesticated dogs spent in the presence of humans shaped their habits and disposition, resulting in an animal that relies on humans for its needs. According to the Humane Society of the United States (HSUS), dogs need humans to provide veterinary care, a nutritionally balanced diet, water, training, exercise, and protection.

A dog that is left outside for any amount of time needs shelter from the elements. During the winter, the HSUS advises to let animals stay indoors during cold weather, but if your dog spends significant time outside, it needs a doghouse that is dry and draft-free. It should be large enough that your dog can comfortably sit in it, but also small enough that it holds the dog's body heat. The house's entrance should not face cold winds, and it should have a covering that allows easy entry but keeps winds out. Insulation, such as straw or cedar bedding, also helps maintain heat. In the summer, a doghouse should provide enough ventilation that your dog can enjoy its shade, but not be suffocated by the heat.

Because dogs are den animals, or animals that like to burrow in enclosed spaces, they are naturally inclined to cozy up in a doghouse. Like other den animals, dogs appreciate having their own space in which to cuddle up and rest that is away from stress and the elements. The size of your dog's house is important; it needs to be the perfect size to feel secure in times of stress — not too big or too small. Your dog will learn that its doghouse is its own personal space, and it will lay inside of it on its own without any coaxing from you. Dogs also take care to keep their space as tidy as they can; an adult dog will rarely use its doghouse as a restroom or purposely soil it. However, a puppy may soil a doghouse if it is too large. The HSUS suggest purchasing a doghouse that will accommodate a dog's adult size, and block off excess space. As the puppy grows, move the barrier back until it needs the full size. Adult dogs need your help to change the bedding; this should be done weekly.

The size of your dog's house also needs to accommodate its quirks. You might notice that your dog likes to circle around a few times before lying down. This is a behavior all dogs exhibit — both domestic and wild. Scientists are not certain why dogs do this, but they do have a few hunches. One theory is that the dog's ancestors did this to trample tall, unkempt grass and rid the ground of unwanted critters hiding there before lying down. Another hypothesis is that wild dogs like to sleep with their noses into the wind, so they turn into circles to find the wind. Either way, it is a dog's natural intuition to do this, and your dog's house should accommodate its natural desires.

Building your own doghouse can be a rewarding and fun experience for the whole family. When your new doghouse is complete, the new home will provide years of enjoyment for your pet and protection from the weather. Depending on the style of doghouse you choose, you can complete the project over the course of one or two weekends. These easy-to-follow plans will simplify the task and provide tips and tricks to help you along the way.

Selecting the Proper Size

Choosing the proper-sized home for your pet is important for the comfort and joy of man's best friend. If the doghouse is too small, it will not be comfortable for your dog, and your furry friend may not even use its new doghouse. Conversely, if the doghouse is too large, it will not provide the secure feeling your dog desires. If you live in a cooler climate, an oversized doghouse will not allow your dog's natural body heat to warm the interior of its new residence. In warmer climates, be sure to provide adequate ventilation and position the doghouse so the front opening faces the predominant direction of the wind.

A good rule of thumb is to select a plan with an interior height that is 20 percent to 30 percent taller than the height of your dog. Be sure that the width of the doghouse is 1 to 2 inches wider than the length of the dog so it can easily turn around when it is inside the house. Measure the length of your dog and add 20 percent to 30 percent to determine the interior length of the doghouse.

The approximate interior dimensions of the doghouse design plans detailed in this book are as follows:

Small doghouse:
16 inches wide by 24 inches long by 18 inches tall

Medium doghouse:
24 inches wide by 28 inches long by 24 inches tall

Large doghouse:
30 inches wide by 36 inches long by 32 inches tall

Extra large doghouse:
36 inches wide by 48 inches long by 44 inches tall

The following represents the size doghouse you should select based on your dog's breed. If you have a mixed breed, try to identify your dog according to the breed that is roughly the same size as your dog.

Small doghouse: Chihuahua, dachshund, Pomeranian, toy poodle, and toy terrier

Medium doghouse: beagle, Boston terrier, cocker spaniel, Jack Russell terrier, shih tzu, and Scottish terrier

Large doghouse: basset hound, chow chow, springer spaniel, husky, and Labrador retriever

Extra large doghouse: Akita, dalmatian, Doberman pinscher, English setter, Old English sheepdog, German shepherd, golden retriever, and rottweiler

This list is intended as a general guide only; be sure to measure your own pet to find the right size doghouse for it. If your dog is just a puppy, research the breed online or check with the breeder that you purchased the pet from to determine the lineage of the dog. If you purchased your dog from a rescue agency or shelter, ask the shelter workers how large the dog will be when it is fully grown.

Building a Doghouse for Multiple Dogs

When constructing a doghouse for multiple dogs, you need to know the demeanor of your individual pets. While dogs are pack animals, your pets may or may not live well together. In many cases, your dog may prefer its own doghouse, so you will need to consider this prior to building your doghouse. Will you need to build small individual doghouses for each pet enclosed in a separate pen? Or can you construct one large home for all to share?

There are several advantages to building individual homes enclosed in separate caged-in areas. First, when it comes time for daily feeding, it is much easier to control the feeding if your dogs are separated. Second, keeping your dogs apart can reduce the spread of disease among your pets. And finally, while your dogs may seem to get along just fine, you never know when a fight may occur. Building separate quarters can ensure the safety of your pets, especially if you have small dogs living with much larger animals.

Choosing a Style

When you have determined the proper size for your pet's new home, you need to choose a design style that complements the architectural style of your home, its surroundings, or at the very least your personal taste. The most common design is a doghouse with a simple hip roof. A hip roof is a common single pitch roof, similar to that of most homes built today. These are easy and cost-effective to build, and they fit in well in most neighborhoods.

If you live in a rural area — for example, on a farm with a barn — and you have other livestock, a more common design is a doghouse with a gambrel roof. This style of roof is similar in appearance to that of an older barn with a common multi-pitch roofline.

To provide your dog with a place to lounge, consider adding an extension to the roof of your doghouse and a front porch.

Selecting Your Materials

The base of your doghouse should be constructed using pressure-treated lumber if it will come in contact with the ground. It is important, however, not to use pressure-treated materials in areas that your pet may come in contact with, such as the interior framing. The main frame can be constructed of standard 2x4 pine lumber available at your local hardware supply store. If you plan to paint your doghouse, use high-quality exterior-grade latex paint and primer.

Site Considerations for Your Doghouse

When choosing a location for your new doghouse, be sure to evaluate the terrain where you plan to build. Do not place the doghouse at the base of a hill where rain can drain into the doghouse. Elevate the doghouse slightly to reduce the chance of flooding.

Do not place the doghouse adjacent to a fence. If you have a large dog, it could jump on the roof of the doghouse, jump over the fence, and escape.

Place the doghouse under a shady tree if you live in an area with a warm climate. Keep the doghouse close to your home to make feeding easy and to provide your pet with a more secure feeling.

Training Your Dog to Like Its New Home

So you spent all weekend building your dog a new home, but it wants nothing to do with the new abode. What should you do?

Because dogs and humans have been interacting for so long, dogs are highly trainable. If your goal is for your dog's house to be a space for it to stay when you are away from the house, or a place for it to sleep at night, you should be able to achieve this easily with proper training. Positive reinforcement and consistency is important in dog training. The first thing you should do is start feeding your dog in its new home. A dog cannot resist food, so if you place its dish just inside the door opening, it will become accustomed to going inside. You can also place its favorite toy inside the doghouse.

If you want your dog to be in its house at night, give it a positive reason to go there when it is bedtime — a treat, perhaps. You could create a simple command for your dog, such as "house" or "lay down." With correct training, your dog will know to get into its doghouse when it hears you say this.

When you first start training your dog to go to its house, be sure to consistently praise your dog with a treat or a toy, in addition to a positive-sounding "Good dog!" When your dog catches on to the command and routine, you can slowly decrease how often your dog gets a treat, but be sure to keep rewarding it occasionally. For more information from the HSUS on training dogs through positive reinforcement, visit **www.humanesociety.org/animals/dogs/tips/dog _training_positive_reinforcement.html**.

Introduce your pet to its doghouse during a moderate season such as spring or fall. Do not expect your dog to sleep outside for the first time on the coldest day of the year. It may take your pet some time to get acclimated to the outdoor weather conditions, hot or cold.

Plan #1 — Medium Size
With Hip Roof:

Bill of Material

Item	Qty	Description	Item	Qty	Description
1	2	2x4 x 2'-4" pressure treated	10	1	2x2 x 1'-9"
2	3	2x4 x 1'-9" pressure treated	11	1	2x2 x 10 7/8"
3	1 box	3" galvanized deck screws	12	2	2x2 x 10"
4	1	1/2" plywood x 2'-0" x 2'-4"	13	2	1/2" plywood x 1'-1 3/4" x 2'-4"
5	1 box	1 5/8" galvanized deck screws	14	2	2x2 x 11 1/4"
6	4	2x2 x 1'-1"	15	3	2x2 x 2'-1"
7	1	2x2 x 1'-10 3/8"	16	4	2x2 x 1'-0 5/8"
8	1	1/2" plywood x 2'-0" x 2'-0"	17	2	1/2" plywood x 1'-6" x 2'-5"
9	1	1/2" plywood x 2'-0" x 2'-0"	18	1 quart	paint

Step 1: Assemble Base Frame:

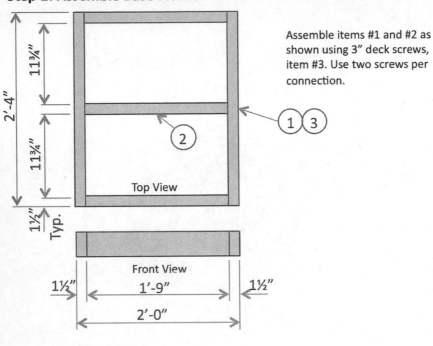

Assemble items #1 and #2 as shown using 3" deck screws, item #3. Use two screws per connection.

Step 2: Install Floor:

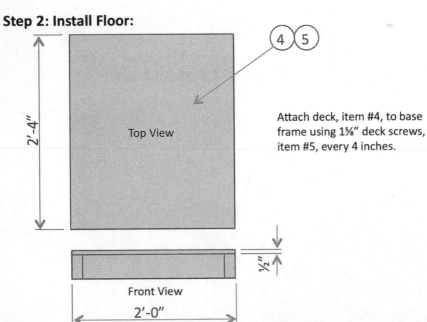

Attach deck, item #4, to base frame using 1⅝" deck screws, item #5, every 4 inches.

Step 3: Assemble Back Wall:

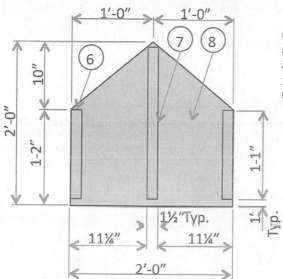

Cut out item #8, back wall, as shown. Attach items #6 and #7 to back wall using 1⅝" deck screws, item #5. Use 4" spacing.

Step 4: Assemble Front Wall:

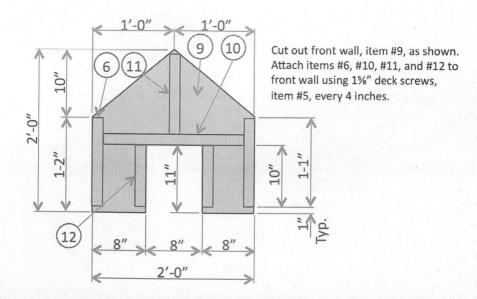

Cut out front wall, item #9, as shown. Attach items #6, #10, #11, and #12 to front wall using 1⅝" deck screws, item #5, every 4 inches.

Step 5: Assemble Side Walls (Qty 2):

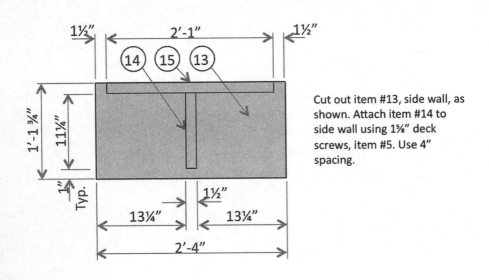

Cut out item #13, side wall, as shown. Attach item #14 to side wall using 1⅝" deck screws, item #5. Use 4" spacing.

Step 6: Attach Side Walls to Base:

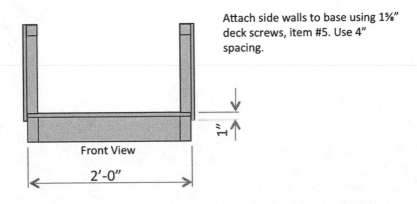

Front View

Attach side walls to base using 1⅝" deck screws, item #5. Use 4" spacing.

Step 7: Attach Rear Wall to Side Walls and Base:

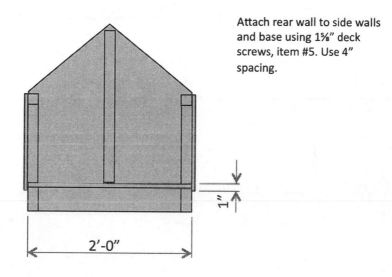

Attach rear wall to side walls and base using 1⅝" deck screws, item #5. Use 4" spacing.

1"

2'-0"

Step 8: Attach Front Wall to Base and Sides:

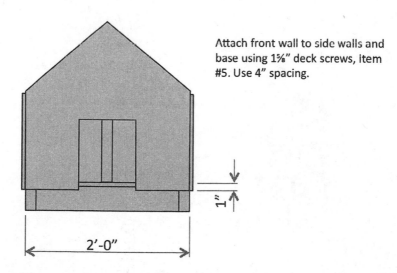

Attach front wall to side walls and base using 1⅝" deck screws, item #5. Use 4" spacing.

1"

2'-0"

Step 9: Assemble Roof and Roof Supports:

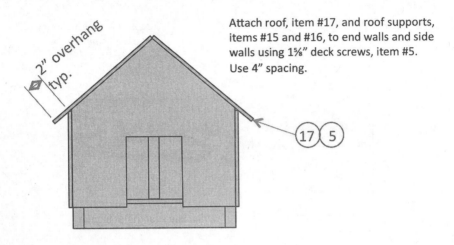

Attach roof, item #17, and roof supports, items #15 and #16, to end walls and side walls using 1⅝" deck screws, item #5. Use 4" spacing.

Step 10: Final Assembly:

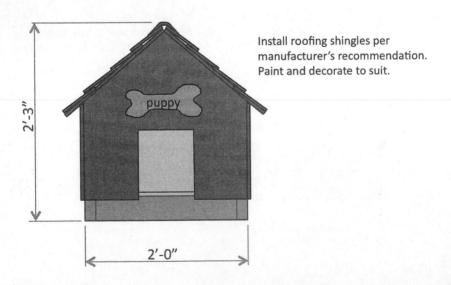

Install roofing shingles per manufacturer's recommendation. Paint and decorate to suit.

Doghouse Plan Variations:

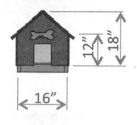

Plan #2 — Small Hip Roof:

Plan #2 is constructed using the same basic design as the detailed plan #1. The overall base dimensions are 16" wide x 24" long. The wall height is 12" with an overall height of 18". The doggie door opening is 8" wide x 8" tall.

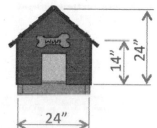

Plan #1 — Medium Hip Roof:

See detailed drawings included in this book.

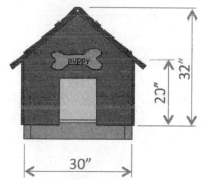

Plan #3 — Large Hip Roof:

Base dimensions are 30" wide x 36" long. The wall height is 20" with an overall height of 32". The doggie door opening is 10" wide x 16" tall.

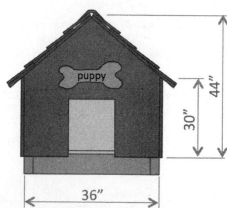

Plan #4 — Extra Large Hip Roof:

Base dimensions are 36" wide x 48" long. The wall height is 30" with an overall height of 44". The doggie door opening is 12" wide x 18" tall.

Gambrel Roof & Wall Modifications for Plan #5, #6, #7, and #8:

Gambrel Roof Step A:

To build a doghouse with a gambrel roof, modify the end walls as shown. Dimensions shown are for a medium doghouse. The roof deck, item #17, will be installed in four pieces.

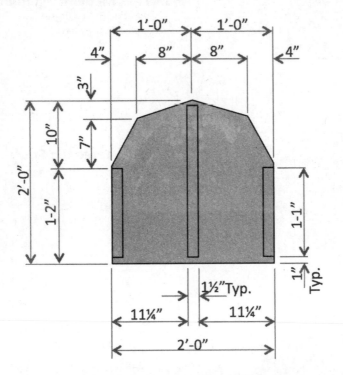

End wall detail for medium doghouse.

Gambrel Roof & Wall Modifications for Plan #5, #6, #7, and #8:

Gambrel Roof Step B:

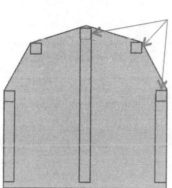

Install 2x2 roof supports as shown. Roof support is the same length as side wall.

Gambrel Roof Step C:

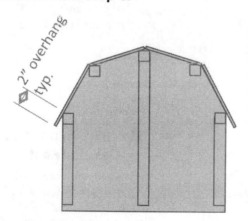

Cut plywood deck to width to match profile of end wall. Attach deck to end wall and 2x2 supports using deck screws. Install shingles per manufacturer's recommendation.

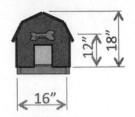

Plan #5 — Small Gambrel Roof:

Plan #5 is constructed using the same basic design as the detailed plan #2. The overall base dimensions are 16" wide x 24" long. The wall height is 12" with an overall height of 18". The doggie door opening is 8" wide x 8" tall.

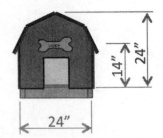

Plan #6 — Medium Gambrel Roof:

Plan #6 is the same overall dimensions as the detailed plan #1. Change the end walls and roof as shown.

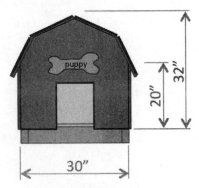

Plan #7 — Large Gambrel Roof:

Base dimensions are 30" wide x 36" long. The wall height is 20" with an overall height of 32". The doggie door opening is 10" wide x 16" tall.

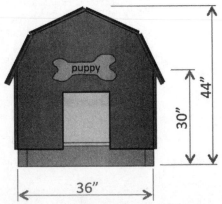

Plan #8 — Extra Large Gambrel Roof:

Base dimensions are 36" wide x 48" long. The wall height is 30" with an overall height of 44". The doggie door opening is 12" wide x 18" tall.

Extended Porch Modifications for Plan #9, #10, #11, and #12:

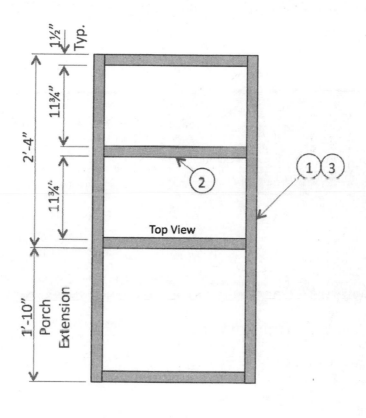

Top View

Front View

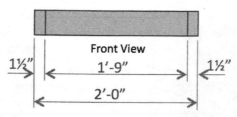

To build a doghouse with a front porch, modify the base as shown. Dimensions shown are for a medium doghouse. Extend the main base frame, item #1, and roof deck, item # 17, by the amount shown. Add two additional vertical supports, item #6, to support the extended roof.

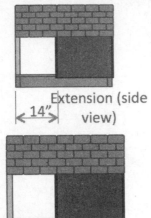

14" Extension (side view)

Plan #9 — Small with Porch:

Plan #9 is similar to plan #2. Extend the roof and base as shown. The overall base dimensions are 16" wide x 38" long.

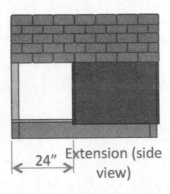

20" Extension (side view)

Plan #10 — Medium with Porch:

Plan #10 is similar to the detailed plan #1. Extend the roof and base as shown. The overall base dimensions are 24" wide x 48" long.

24" Extension (side view)

Plan #11 — Large with Porch:

Plan #11 is similar to plan #3. Extend the roof and base as shown. The overall base dimensions are 30" wide x 60" long.

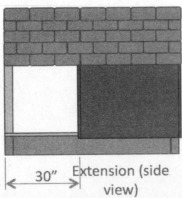

30" Extension (side view)

Plan #12 — Extra Large with Porch:

Plan #12 is similar to plan #4. Extend the roof and base as shown. The overall base dimensions are 36" wide x 78" long.

A HOME
FOR YOUR CAT

Cats are one of the quirkiest pets. Cat lovers appreciate that their cats can snuggle up for hours, but still maintain the "cattitude" that makes them highly indepe ndent and unpredictable. People like having cats in and around the house because they are a delight to watch, and their soft, clean fur is silky to pet and cuddle with. People like that cats are relatively low-maintenance pets. A cat will clean itself, help itself to the litter box, and can entertain itself with a ball, a piece of string, or simply a window.

Throughout history, cats have been one of the most highly regarded animals because of their beauty, grace, and poised demeanor. It is thought the Egyptians domesticated cats around 2,000 B.C., and today's house cats are the descendents of ancient Egypt's first domesticated cats. It took the Egyptians more than 500 years to domesticate the cat, but when done, cats not only became pets, but were also used to hunt fowl and fish. They became a sacred animal to the Egyptians. Cat amulets and pendants were thought to bring protection, and cat statues were used in shrines and in Egyptian burials.

Today, cats rule the house; you cannot make a cat do something it does not want to do, which is one reason why people love these pets. As pets that are not totally "trainable," cats tend to walk around their owners' houses like they own them. Cats are territorial animals, and they consider your house to be theirs. They will mark your house as their territory, leaving visible signs, if not provided with the correct amenities. A properly built cat tree house that has the correct features will provide your cat with a space to spend hours of time without taking a toll on your home.

For example, you may notice your cat likes to claw at furniture or carpet, leaving behind tears or snags. Even if your cat is declawed, it will still scratch at rough surfaces. There are several reasons cats like to scratch. Scratching is one way cats mark your house as their territory. In addition to the visible mark your cat's claws leave on your house, scratching also leaves behind your cat's scent, as claws have scent glands. Scratching also provides cats with exercise, and it feels good and sharpens their claws.

Although you cannot train your cat to do anything it does not want to do, there are ways to entice your cat to not scratch furniture. Including a scratching post in your cat tree house plan is one way to save your upholstery while still allowing your cat to scratch. A scratching post should be covered in a material that is good for a cat's claws, such as carpeting or upholstery. The post should be sturdy enough that a cat can push on it without it falling down, and it should also be tall enough for a cat to fully stretch out while scratching. Wait as long as possible before re-covering a scratching post because a cat's scent is all over its post, and cats prefer clawing shredded materials.

Encourage your cat to use its scratching post by scenting it with catnip and hanging toys off it. Some cat owners simply place catnip leaves around their cat's scratching post to entice their cat to the area. However, if you do not want to end up with leaves strewn around the house, there are several non-toxic catnip sprays on the market made from catnip oils, and they will draw your cat to toys or posts that you spray. Do not try to make your cat use the scratching post by forcing it to run its claws on it; this will only scare the cat, and it might scratch you.

Some cat owners try to deter their cat from scratching household items by placing tape over problem areas, or by spraying their cat's favorite scratching areas with cologne or a citrus scent, which most cats will avoid. Other cat own-ers deter their cats by startling them with clapping or using a squirt of water when they catch their cat in the act. The American Society for the Prevention of Cruelty to Animals

(ASPCA) advises using clapping and water squirting only as a last resort. If scare tactics, such as water squirting, are used too often, a cat will begin to associate this frightening behavior with you, rather than the behavior you are trying to stop. Your cat will begin avoiding you at all costs and will still scratch at your furniture — a no-win situation.

Besides scratching, you may notice your cat likes to perch on high spots like your countertops — or any high spot it can possibly get to. A cat's feet paw around in the litter box and walk on the floor, so letting your cat jump on the counter where you prepare meals is unsanitary. Jumping and liking high places is in a cat's natural instincts. In the wild, cats climb trees to hunt, and domesticated cats still have these instincts. Cats are built to jump with strong back muscles and hindquarters, and their claws help them grab onto surfaces when climbing.

A cat's tendency to climb as high as possible also comes from its territorial nature — the higher a cat can get, the better it can see its territory. Height also provides cats a place to retreat when they are scared or trying to avoid someone or something. You can encourage your cat to use its tree house as a perch by enticing it with some catnip or a toy. The ASPCA advises precariously setting baking sheets on the edge of countertops if you do not want your cat jumping there; when the sheets move or fall, the movement and noise will scare your cat. Do not scold your cat or push it off the counter. This will only make the cat scared of you, rather than the counter.

House cats are known for their ability to sleep throughout the entire day. Cats especially like snuggling up in small, confined spaces. Cat owners are always amazed at the contortionist-like ways cats can fit into tiny spaces. This behavior comes from a cat's natural desire to feel warm and protected. Cats tend to sleep with one eye open and both ears perked because of their instinct to be alert to dangers. A snug, tight place gives a house cat a place where it feels safe and protected from danger. A cat tree house can be built to accommodate this. Your cat will feel secure if it has a space that is enclosed on three sides.

When your cat is not eating, sleeping, or scratching, it is likely playing. Cats can turn virtually anything into a toy. Anything that dangles, rolls, slides, or moves will catch your cat's attention. As natural hunters, a cat's instinct is to stalk and chase its prey; for a house cat, this instinct translates into stalking and chasing toys or moving objects. If you sit on the couch and let your hand dangle off the end, do not be surprised if your cat attacks your hand. This "hunting" is important for your cat because it is how it gets exercise, builds muscle, and stays mentally stimulated. A cat tree house that has a dangling toy will automatically entice your cat and provide hours of entertainment.

Building a home for your house cat will provide a place where your cat can lounge and claw, which will help deter it from clawing at your furniture. Place your cat's new home near its favorite hangout, and your cat will acclimate to its new home more quickly. You can even install a toy in your cat's home to occupy it. Your cat may also use its new home as a place to retreat from small children or escape from other pets in your home.

Choosing a Plan

For young, healthy cats, choose a plan with several elevations that your cat can leap to. As your cat gets older, you may want to consider a lower-profile plan with ramps to prevent your cat from having to leap great distances. The plan variations detailed in this book include two versions of most plans: one for typical-sized cats and a larger version for larger cats or fat cats.

Selecting your Materials

Here are the materials you will need to construct one of the cat houses in this chapter:

Wood: Pine is the preferred type of wood if you intend to cover your project with carpet. Pine is easy to work with and is readily available. If the wood will be exposed, cedar is excellent softwood to work with, and it is durable if exposed to the weather. In the areas of your cat's home where the wood will be exposed, choose softwood so your cat can sink its claws into the post as it scratches. Some common softwood species include southern yellow pine, red cedar, and spruce.

Carpet: Use a non-looped carpet so your cat can scratch without getting its claws tangled. Using plush carpet would be a good choice because the tops of the loops in this kind of carpet are cut off to create an even finish. Berber and other expensive woven carpets are manufactured with long strands and closed loops, and your pet may get its claws caught in the loops. Visit your local carpet retailer to find the right material; you may even find an appropriate fabric with a reduced price that is left over from a large installation project.

Rope: Cover the scratching post with sisal rope, which is similar to twine. As an alternative, sisal fabric is also available, and some cats may prefer the fabric over the rope. The fabric can be attached to the structure in full-sized sheets, while the rope must be wrapped around the scratching post.

Selecting a Cat Home for Multiple Pets

Cats are territorial animals and tend to mark their territory to establish ownership. If you build a small cat house for multiple cats, fighting among the cats will most likely ensue. To avoid cat fights, construct a large home with multiple levels and multiple enclosures. Another option is to build several smaller units and place them in different areas throughout your home.

Cats also tend to be more active at night, so you may want to place the cat home in a common area such as a family room. If possible, isolate this area from bedrooms so your cat will not disturb your sleep.

Construction Tips

Before you begin constructing a new home for your feline companion, here are some tips to make building a new home much easier:

- Attach the carpet and sisal rope with high-quality glue and carpet tacks or heavy-duty staples.

- Be sure the glue has dried 100 percent before you allow your cat access to its new home.

- Pre-drill undersized holes in wood prior to inserting screws to prevent the wood from splitting.

Cat Tree House Plans
Plan #13 — 24 x 28 Three-Post Plan:

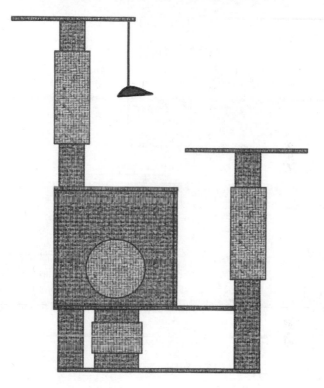

Bill of Material

Item	Qty	Description	Item	Qty	Description
1	1	4 x 4 x 4'-0"	10	1	1/2" plywood x 1'-5" x 1'-5"
2	2	4 x 4 x 8"	11	1	1/2" plywood x 1'-4 1/2" x 1'-4 1/2"
3	1	1/2" plywood x 2'-0" x 2'-4"	12	1	1/2" plywood x 1'-4 1/2" x 1'-5"
4	1	4 x 4 x 2'-6"	13	2	1/2" plywood x 1'-4" x 1'-4"
5	1 box	3" deck screws	14	1	1/2" plywood x 6" x 11"
6	1	1/2" plywood x 2'-0" x 2'-4"	15	1 roll	carpet
7	1 box	1 5/8" deck screws	16	1 yd	sisal carpet
8	1	1/2" plywood x 1'-4" x 1'-4"	17	1	toy
9	1	1/2" plywood x 1'-4" x 1'-4 1/2"	18	1	string

Step 1: Assemble Base Frame:

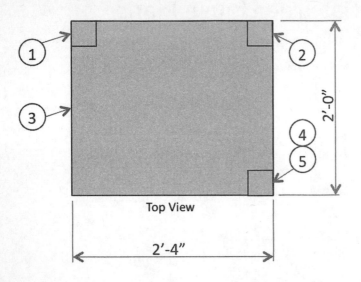

Top View

2'-0"

2'-4"

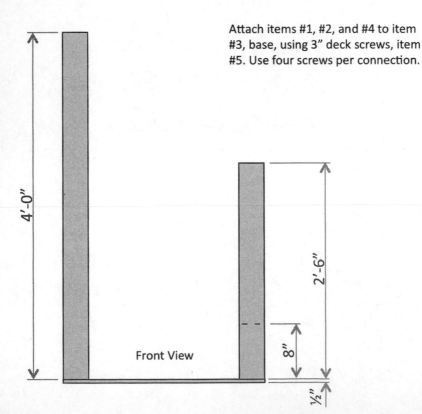

Attach items #1, #2, and #4 to item #3, base, using 3" deck screws, item #5. Use four screws per connection.

4'-0"

2'-6"

8"

½"

Front View

Step 2: Assemble Tier 2 Floor:

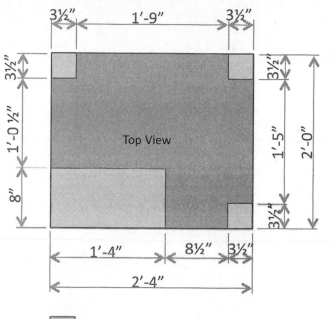

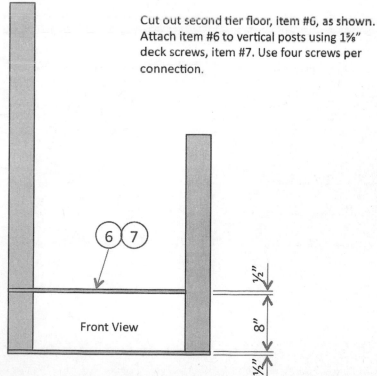

Cut out second tier floor, item #6, as shown. Attach item #6 to vertical posts using 1⅝" deck screws, item #7. Use four screws per connection.

Step 3: Cut Out House Components:

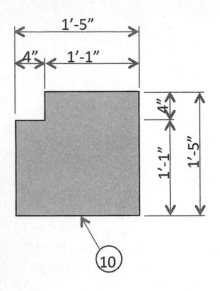

Cut out items #8, #9, and #10 as shown.

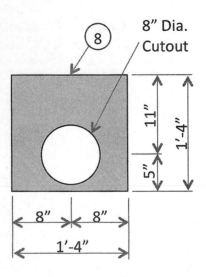

8" Dia. Cutout

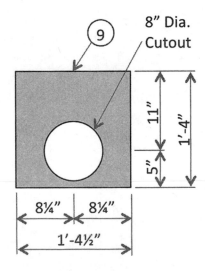

8" Dia. Cutout

Step 4: Assemble House:

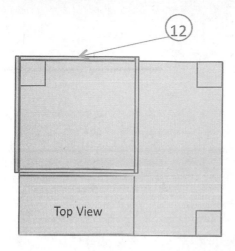

Top View

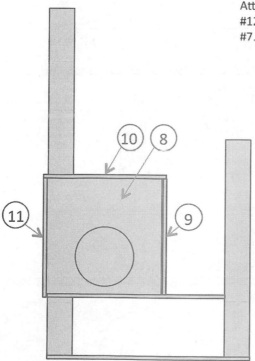

Attach items #8, #9, #10, #11, and #12 as shown using 1⅝" screws, item #7. Use 4" spacing.

Front View

Step 5: Assemble Ramp and Upper Tiers:

Attach ramp, item #14, to house and base using 1⅝" deck screws, item #7. Cut out item #13, upper tiers, and attach to top of post using 3" deck screws, item #5. Use four screws per connection.

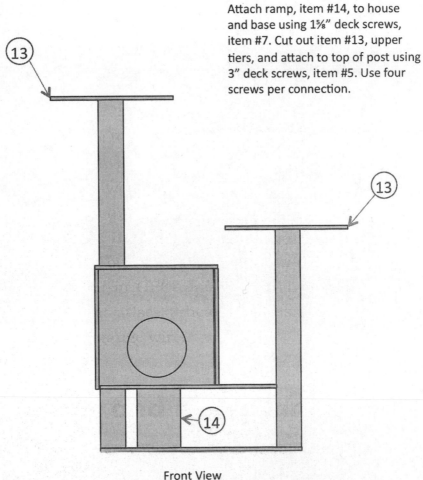

Front View

Step 6: Final Assembly:

Cover entire surface area with plush carpet, item #15, using staples or glue. Attach scratching carpet, item #16, to upper posts and ramp. Attach item #17, toy, to upper tier using item #18, string.

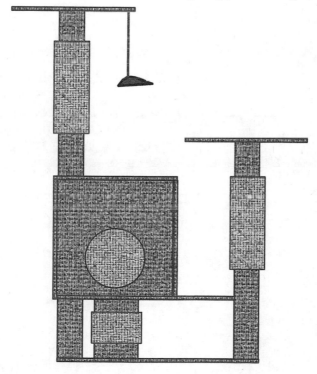

Cat Tree House Plan Variations:

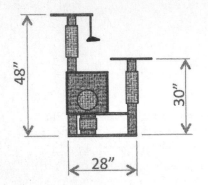

Plan #13 — 24 x 28 Three-Post Plan:

See detailed drawings
included in this book.

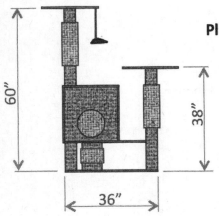

Plan #14 — 30 x 36 Three-Post Plan:

Same design as plan #13 for larger cats.
Base is 30" x 36". Enclosure is 20" wide x
20" high x 20" long. Upper tower is 60"
high. Lower tower is 38" high.

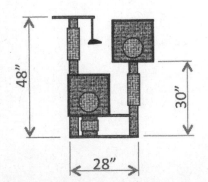

Plan #15 — 24 x 28 Three-Post Plan with Double Cat House:

Same basic design as plan #13. Add
second cat house on middle tier.

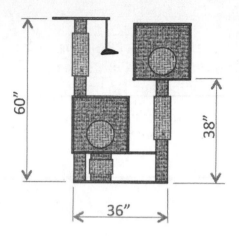

Plan #16 — 30 x 36 Three-Post Plan with Double Cat House for Larger Cats:

Same basic design as plan #14 for larger cats. Add second cat house on middle tier.

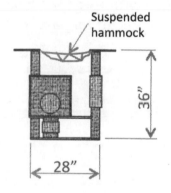

Suspended hammock

Plan #17 — 24 x 28 Three-Post Plan with Hammock:

Similar to plan #13. Both posts are 36" tall. Upper landings are 10" x 16". Suspend a hammock between the posts.

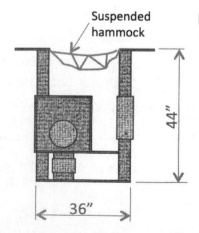

Suspended hammock

Plan #18 — 30 x 36 Three-Post Plan with Hammock for Larger Cats:

Similar to plan #14. Both posts are 44" tall. Upper landings are 10" x 16". Suspend a hammock between the posts.

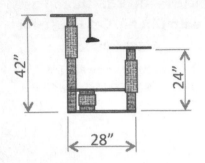

Plan #19 — 24 x 28 Simple Three-Post Plan with Three Landings and No Box:

Simple plan similar to plan #13 with three landings and the box removed.

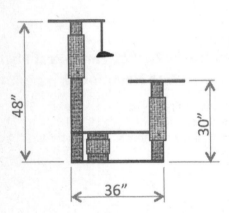

Plan #20 — 30 x 36 Simple Three-Post Plan with Three Landings and No Box for Larger Cats:

Simple plan similar to plan #14 with three landings and box removed.

Internal connecting cutout

Roof access cutout

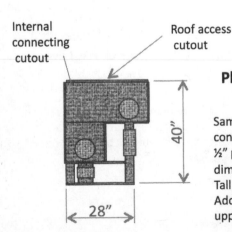

Plan #21 — 24 x 28 Three-Post Plan with Large Upper Cat House:

Same basic design as plan #13. Add large connecting cat house above lower house. Use ½" plywood to construct upper box. Overall dimensions of upper box are 16" x 16" x 34". Tall post is 24" high. Short posts are 8" high. Add an 8" diameter interior cutout to connect upper and lower boxes.

Plan #22 — 30 x 36 Three-Post Plan with Large Upper Cat House for Larger Cats:

Same basic design as plan #14. Add large connecting cat house above lower house. Use additional ½" plywood to fabricate upper box. Overall dimensions of upper box are 20" x 20" x 42". Tall post is 30" high. Short posts are 8" high. Add an 8" diameter interior cutout to connect upper and lower boxes.

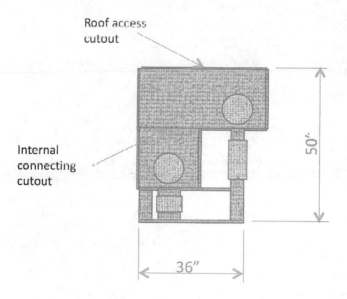

Roof access cutout

Internal connecting cutout

50'

36"

Chapter 4

BIRDCAGES, BIRDHOUSES, AND FLIGHT ENCLOSURES

People love watching birds because they are so unique. Many enjoy keeping birds as pets because they are social and interactive, do not take up much space, and are relatively low-maintenance compared to other pets. Most do not think of birds as pets that you can play with or use for companionship, but this is far from the truth. Birds are highly trainable pets, which is why so many zoos have impressive bird shows. Birds can sit with you for hours, they can be playful, and some birds can even be trained to "talk."

The Egyptians were the first people to have birds as pets more than 4,000 years ago. In ancient times, watching pet birds was the equivalent of watching television today. Birds were often gifts for the elite. It is said Alexander the Great was granted a parakeet for invading Northern India, and Christopher Columbus gave Queen Isabella of Spain a pair of Cuban Amazon parrots from the New World. For more practical purposes, miners used canaries to help detect poisonous gasses, and pigeons were used throughout military history as a way to send messages.

In the wild, birds have infinite space to fly and move around, so it is important for domesticated birds to have sufficient space. The ASPCA suggests providing your pet bird with the largest cage you can afford and can accommodate in your home or on your property. Your bird should be able to stretch its wings and fly a short distance in its cage. Birds kept inside the house should be allowed to be outside their cage for exercise; the length and frequency of exercise outside the cage depends on your bird's breed. A major bonus of keeping your pet birds in an outdoor aviary is that these can often be built large enough that your bird always has sufficient room for exercise. However, if your birds live inside, let them explore an enclosed room for exercise; just be sure to cover any mirrors or windows to prevent your bird from flying into them.

Any structure built for birds should include perches. Birds perch up high in the wild for a variety of reasons. Perching is how birds rest, and it also provides birds with visibility so they can see predators. Also, perching defends birds from predators on the ground. You should include perches of varying heights for your pet bird, taking into consideration the size of its feet. The wrong perch size can lead to arthritis and soreness. It is recommended that a bird's perch allow a bird's foot to wrap 75 percent around the perch.

According to Dr. Greg Rich, a Louisiana avian veterinarian, you should provide your bird with three different perches. One of these perches is for sleeping, and the other two should be made of different materials and size to avoid boredom. Including perches of different materials and sizes also makes your bird's cage more like its natural environment, where birds find perches of all sizes and shapes. Be sure perches are not over your bird's water and food — you do not want droppings to contaminate them.

If you ever watch a bird in the wild, you will notice it is constantly moving. Birds hardly ever sit still; they continuously look around, peck, and chirp. Pet birds will become bored if not stimulated — some more than others, depending on breed. To avoid boredom, be sure your bird's cage has toys. The best toys allow your bird to climb, chew, hide, or manipulate the toy in some way. Bird toys could include ladders, swings, bells, and wooden chew toys. Chewing toys are also important for keeping a bird's beak trimmed. Many birds enjoy having a mirror in their cage as well. The size and type of toys your bird needs depends on its breed and size. Most store-bought toys are labeled to identify what breeds they are appropriate for.

When setting up your pet bird's home, it will be a process of trial and error, so you need to be flexible to change to accommodate its needs. In the wild, birds are prey animals. Even though your bird is in no danger of being eaten, it will still scare easily. Things like the shadow of a wild bird outside, noises, other animals, strange people or objects, and the dark are just a few of the things that can scare your bird. You might have to experiment with different toys as some may frighten your bird.

Some people prefer to watch birds from a distance in the wild, rather than keeping them as pets. Having a birdhouse in your backyard is a good way to watch and enjoy birds without having the responsibility of feeding and entertaining them. In North America, there are about 36 bird species that will nest in a birdhouse, and you can build houses with features to attract specific bird species. It is best to install birdhouses in the winter and early spring because this is when birds look for nesting spots in the wild.

Birdhouses should accommodate the birds in your area. You can attract certain bird species with boxes that have the right size hole. The height at which you place your birdhouses should also be a factor, as well as how far apart you place the houses. For example, a house for bluebirds should be 5 or 6 feet tall and 300 feet from any other birdhouse. Unlike domesticated birds you keep in your house, nesting birds in the wild do not need a perch. A perch will only cause problems for the birds because predators will use the perch to get inside to the nest.

You can attract birds to your property and into your birdhouse by making sure your property has all the amenities birds need. You can include trees, shrubs, and plants that produce fruit or seeds in your landscaping. It is good to have some plants that hold seeds and fruit through the cold winter months. Besides food, birds need water. A birdbath is not only aesthetically pleasing, but it will help you keep birds on your property and in your birdhouses. Birds need water for both drinking and bathing; the best sources of water for birds are between 1 and 3 inches deep.

There are three distinct types of homes for birds: birdcages, birdhouses, and flight enclosures.

Basic birdcages are designed for indoor use for your small pet birds. Birdhouses are constructed for outdoor use as a habitat for wild birds. Placing a birdhouse near your home, deck, or gazebo can provide hours of bird-watching enjoyment for bird lovers.

Flight enclosures consist of a large outdoor area to house birds of prey. The large cage or fenced-in area provides a secure location for your pet bird of prey to exercise while being protected from the dangers of predators. Typical birds of prey include eagles, falcons, hawks, osprey, and owls. Other not-so-common birds of prey include harriers, kites, and merlin.

Birdcage Design

While some indoor birds prefer to have freedom to explore your home, it is best to provide a cage for nighttime sleeping and a place for feeding. Typical indoor bird species kept as pets include budgies, canaries, cockatiels, cockatoos, finches, lovebirds, macaws, parakeets, parrots, and toucans.

When constructing your cage, be sure to use stainless steel wire for durability and for the safety of your pet. Birds tend to gnaw on the cage wire, and if you use painted wire, they may eat the paint. Cage size and style can vary based on the type of bird and number of birds housed in the cage. Follow these guidelines as a general rule for determining the size of the wire diameter of the cage, wire spacing, and approximate cage dimensions.

Small Birds — budgies, canaries, finches, and lovebirds:

- Wire size: 3/32-inch to 1/8-inch diameter

- Wire spacing: 3/8 inch

- Cage dimensions: 18 inches wide by 18 inches deep by 24 inches tall

Medium Birds — cockatiels and parakeets:

- Wire size: 3/16-inch diameter

- Wire spacing: ½ inch

- Cage dimensions: 24 inches wide by 24 inches deep by 36 inches tall

Large Birds — cockatoos, macaws, parrots, and toucans:

- Wire size: ¼-inch diameter

- Wire spacing: 1 inch

- Cage dimensions: 48 inches wide by 30 inches deep by 60 inches tall

When designing your birdcage, create a removable false floor for easy clean up. This removable floor should be constructed so it can slide out from the side so you do not have to dismantle the entire cage every time you clean it.

Be sure to include several perches, toys, an easy-to-access feeder, and a water source. Popular bird toys include mirrors and colorful rope chew toys. You should also make or buy a cover for your birdcage. If you decide to make a cover for your birdcage, split one side panel and add a fabric adhesive strip to make it easier to remove the cover. The length and width of the cover should be three quarters of an inch larger than the cage itself.

The purpose of the birdcage cover is to eliminate the cool air draft and to help keep your pet warm at night. Place the cover over the cage at about the same time every night to encourage your bird to form the same sleeping habits that you have.

Birdhouse Design

By constructing the proper size and style of birdhouse and placing it in the correct location, bird watchers can enjoy the beautiful wild birds in their own backyard. Common species that might take advantage of your homemade birdhouse are bluebirds, cardinals, chickadees, owls, purple finches, purple martins, robins, sparrows, starlings, swallows, wrens, and woodpeckers.

Refer to the chart below for typical design dimensions by species, including the critical entrance hole size and location.

Birdhouse Dimensions						
Species	L	W	H	Entrance Hole Diameter	Hole Height	Mounting Height
Bluebird	6"	6"	12"	1½"	6"	6' to 10'
Chickadee	5"	5"	9"	1-1/8"	6"	6' to 10'
Purple Martin	7"	7"	6"	2-1/8"	1½"	10' to 20'
Owl	8"	9"	12"	3"	8"	10' to 20'
Sparrow	4"	4"	8"	1¼"	6"	10' to 20'
Swallow	5"	5"	6"	1½"	3"	10' to 15'
Titmouse	4"	4"	10"	1¼"	5"	6' to 15'
Robin	8"	8"	8"	open front	open front	6' to 15'

Birdhouse Dimensions						
Species	L	W	H	Entrance Hole Diameter	Hole Height	Mounting Height
Wood-pecker	4"	4"	16"	1¼"	6"	6' to 20'
Wren	4"	4"	10"	1-1/8"	8"	6' to 10'

Bird of Prey Enclosure Design

With their large wing spans and tremendous flight speeds, birds of prey require a large area for flight. As a minimum for smaller birds of prey such as small owls and hawks, the enclosure should measure 10 feet wide by 20 feet long by 10 feet tall. For birds with larger wingspans and higher flight speeds like vultures, eagles, falcons, and gray owls, a much larger enclosure is required. A typical enclosure for these species may be as large as 24 feet wide by 100 feet long by 20 feet tall. At a minimum, you should design your enclosure to be 20 feet wide by 80 feet long by 18 feet tall.

The structural frame of the enclosure can be constructed using standard dimensional lumber. Select standard 4x4, 6x6, or 8x8 wood beams for the posts and 2x4 or 2x6 for the intermediate supports. Fasten the entire structure together using 3/8-inch diameter lag screws for strength and longevity of the enclosure. Lag screws are a heavy-duty screw available at your local hardware store.

Cover the framework with a stainless steel wire mesh with a minimum of 16 gauge wire. To keep predators out and your treasured birds of prey in, select a mesh with a small opening size of about 1 inch by 1 inch. You can substitute galvanized wire mesh as a more economical alternative to stainless steel wire. If you are using galvanized wire, for the safety of your birds, be sure to remove any loose zinc coating prior to initial use.

Cleaning Your Bird Cage

While birds can be an enjoyable pet and provide hours of entertainment, they can also create a large mess. Bird droppings left unattended can attract insects and pests. To ensure your pet's good health, here a few quick cleanup and mess-prevention tips:

- Place a liner on the bottom of the cage. Replace the liner daily either just before bedtime or early each morning.

- Empty bird food cups daily.

- Remove wet bird food as soon as you can so the food does not rot.

- Place a skirt around the bottom of the cage to contain flying debris.

- Place a mat under the cage that is several inches larger than the cage itself. The mat will catch any debris that may fall from the cage.

Feeding Your Pet Bird

It is important to keep a fresh supply of water on hand for your pet bird. Water left unattended for long periods can grow bacteria and fungi, which will make your bird ill.

Appropriate food for pet birds varies greatly by species. Be sure to research the proper portions and type of food that is suitable for your specific pet. To keep food fresh, you can store pellets in the freezer. For fresh-cut food such as fruits and vegetables, precut the food and store it in a plastic bag in your refrigerator. Discard food that is more than five days old.

Here are a few tips for feeding wild birds in your backyard:

- Place the nest boxes and bird feeders close to trees so the birds will feel like they are in their natural environment.

- Place a water bowl or birdbath in the yard near the bird feeder to attract more birds.

- Keep the feeders and nest boxes out of reach from your pet dog or cat.

- Place the birdhouse and bird feeder in the line of sight from your kitchen or dining room window so you can enjoy the birds each day.

Toys for Your Pet Bird

Just like when selecting safe toys for your children who are less than 3 years old, you need to make the same considerations when purchasing toys for your pet bird. Some toys with small parts can create a choking hazard for your pet, while in other cases the birds' claws or beak may become tangled in small openings or mesh. Today, many bird toy manufacturers provide a label stating what bird species the toys are suitable for.

Here are a few tips for purchasing bird toys and introducing them to your pet:

- Remove small parts such as the clapper in a bell or wire hooks, as these may become choking hazards for your pet.

- Look for potential hazards such as toys with holes that may catch your birds' toes or beaks.

- Remove rope toys from the cage after they have become worn and frayed so your pet does not become tangled in the loose strands of fabric.

- Observe your pet when you introduce a new toy to ensure it can use the toy safely without becoming entangled.

Birdhouse Plans

Plan #23 — Purple Martin Birdhouse:

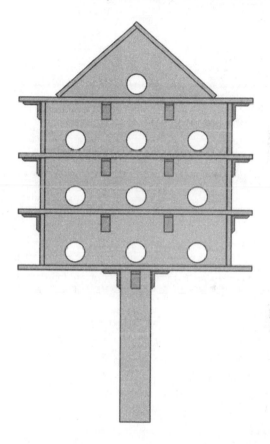

Bill of Material

Item	Qty	Description	Item	Qty	Description
1	3	1/2" plywood x 2'-3" x 2'-3"	9	12	1/2" plywood x 6" x 1'-9"
2	1	1/2" plywood x 2'-3" x 2'-3"	10	1 box	15/8" deck screws
3	2	1/2" plywood x 8" x 1'-6"	11	28	2 x 2 angle braces
4	1	1/2" plywood x 8" x 1'-6"	12	1	4x4 pole x 20'-0"
5	1	1/2" plywood x 1'-0" x 1'-6"	13	1 bag	ready-mix concrete
6	1	1/2" plywood x 1'-0 1/2" x 1'-6"	14	1 quart	primer
7	6	1/2" plywood x 6" x 1'-10"	15	1 quart	paint
8	6	1/2" plywood x 6" x 1'-9"			(white is the color of choice)

Step 1: Cut Out Floor Components:

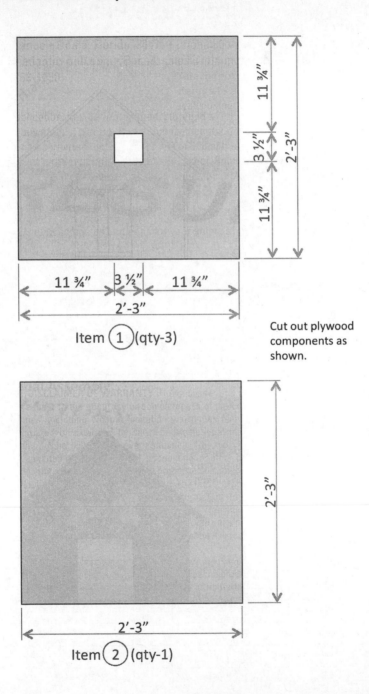

Item ① (qty-3)

Cut out plywood components as shown.

Item ② (qty-1)

Step 2: Cut Out Roof & Divider Components:

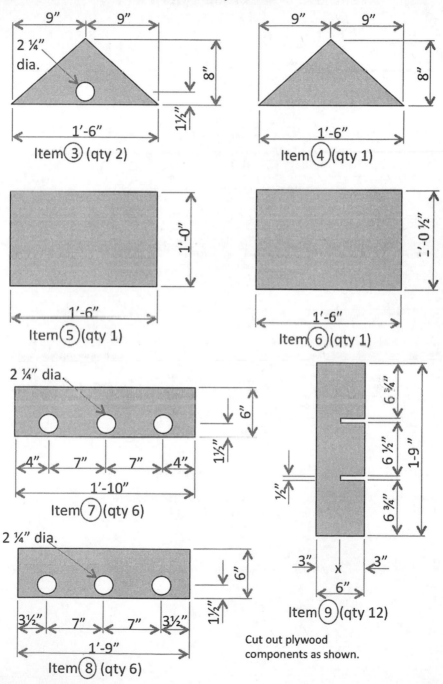

Item ③ (qty 2)

Item ④ (qty 1)

Item ⑤ (qty 1)

Item ⑥ (qty 1)

Item ⑦ (qty 6)

Item ⑧ (qty 6)

Item ⑨ (qty 12)

Cut out plywood components as shown.

Step 3: Assemble Floor and Dividers (qty-3 sets):

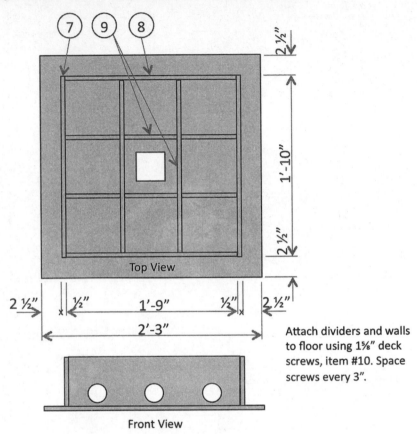

Top View

Front View

Attach dividers and walls to floor using 1⅝" deck screws, item #10. Space screws every 3".

Step 4: Attach 3 Floor Assemblies Together:

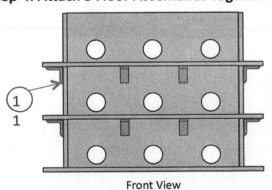

Front View

Attach three floor assemblies together using angle clips, item #11, as shown.

Step 5: Assemble Roof:

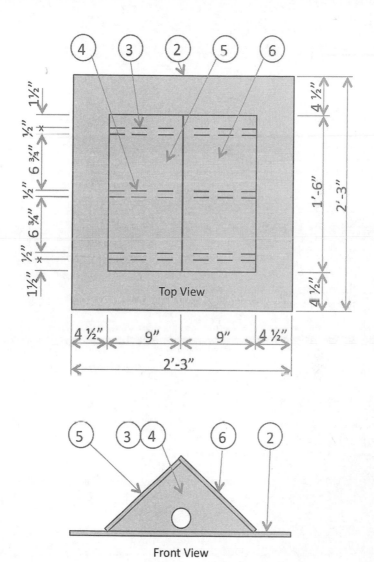

Top View

Front View

Assemble roof and penthouse using 1⅝" deck screws, item #10. Space screws every 3".

Step 6: Attach Roof to Main Assembly and Mount to Pole:

Attach roof and penthouse to main birdhouse assembly using 1⅝" deck screws, item #10. Space screws every 3".

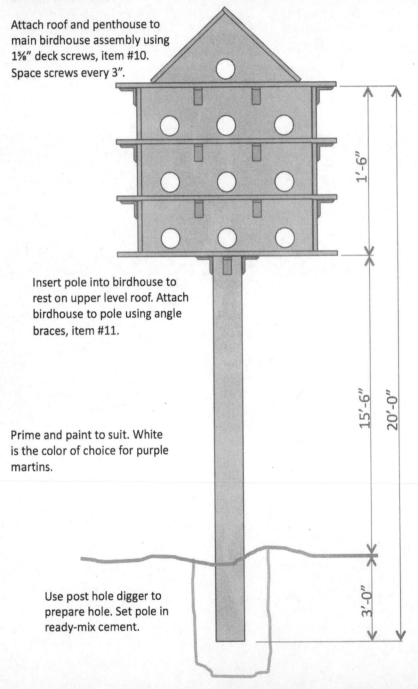

Insert pole into birdhouse to rest on upper level roof. Attach birdhouse to pole using angle braces, item #11.

Prime and paint to suit. White is the color of choice for purple martins.

Use post hole digger to prepare hole. Set pole in ready-mix cement.

1'-6"

15'-6"

20'-0"

3'-0"

Birdhouse Plan Variations:

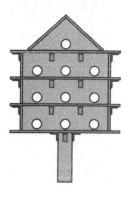

Plan #23 — Medium Purple Martin Birdhouse:

See detailed drawings and bill of material included in this book.

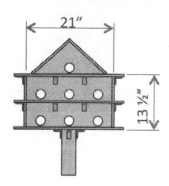

Plan #24 — Small Purple Martin Birdhouse:

Similar to plan #23 designed with two layers. See detailed bill of material for quantities and size of components.

Plan #24 Bill of Material

Item	Qty	Description	Item	Qty	Description
1	2	1/2" plywood x 2'-3" x 2'-3"	9	8	1/2" plywood x 6" x 1'-9"
2	1	1/2" plywood x 2'-3" x 2'-3"	10	1 box	1 5/8" deck screws
3	2	1/2" plywood x 8" x 1'-6"	11	28	2 x 2 angle braces
4	1	1/2" plywood x 8" x 1'-6"	12	1	4x4 pole x 20'-0"
5	1	1/2" plywood x 1'-0" x 1'-6"	13	1 bag	ready-mix concrete
6	1	1/2" plywood x 1'-0 1/2" x 1'-6"	14	1 quart	primer
7	4	1/2" plywood x 6" x 1'-10"	15	1 quart	paint
8	4	1/2" plywood x 6" x 1'-9"			(white is the color of choice)

Birdhouse Plan Variations:

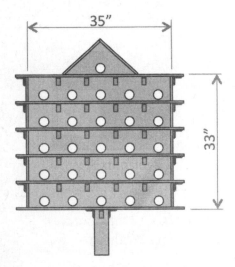

Plan #25 — Large Purple Martin Birdhouse:

Similar to plan #23 designed with five layers and five individual units per side. See detailed bill of material for quantities and size of components.

Plan #25 Bill of Material

Item	Qty	Description	Item	Qty	Description
1	5	1/2" plywood x 3'-5" x 3'-5"	9	40	1/2" plywood x 6" x 2'-11"
2	1	1/2" plywood x 3'-5" x 3'-5"	10	1 box	1 5/8" deck screws
3	2	1/2" plywood x 8" x 1'-6"	11	28	2 x 2 angle braces
4	1	1/2" plywood x 8" x 1'-6"	12	1	4x4 pole x 20'-0"
5	1	1/2" plywood x 1'-0" x 1'-6"	13	1 bag	ready-mix concrete
6	1	1/2" plywood x 1'-0 1/2" x 1'-6"	14	1 quart	primer
7	10	1/2" plywood x 6" x 3' 0"	15	1 quart	paint
8	10	1/2" plywood x 6" x 2'-11"			(white is the color of choice)

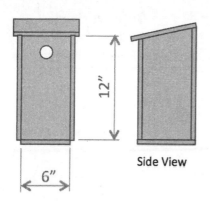

Side View

Plan #26 — Bluebird Birdhouse:

Bluebird birdhouse is 6" wide x 6" deep x 12" tall with a 1½" diameter entrance hole. Use ½" plywood to construct birdhouse. Bottom board is 6" x 6". Top board is 8" x 8". Front wall is 6" x 12". Rear wall is 6" x 13". Side walls are 7" x 13" tapered at the top.

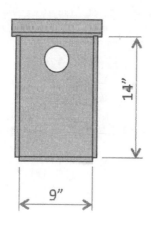

Plan #27 — Owl Birdhouse:

Owl birdhouse is 9" wide x 9" deep x 14" tall with a 3" diameter entrance hole. Use ½" plywood to construct birdhouse. Bottom board is 9" x 9". Top board is 10" x 10". Front wall is 9" x 14". Rear wall is 9" x 15". Side walls are 10" x 15" tapered at the top.

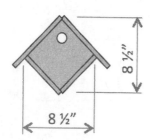

Plan #28 — Wren and Chickadee Birdhouse:

Wren and chickadee birdhouse is 8½" wide x 8½" deep x 8½" tall with a 1⅛" diameter entrance hole. Use ½" plywood to construct birdhouse. Front wall and back wall are 6" x 6". Bottom boards are 6" x 8". Roof boards are 8" x 8".

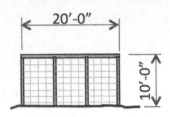

Plan #29 — Outdoor Hawk Enclosure:

Hawk enclosure is 10' wide x 10' tall x 20' long.
Step 1: Place 4x4 x 13'-0" posts in 3'-0" hole set in concrete.
Step 2: Install 2x4 mid supports at top and bottom of posts.
Step 3: Attach wire mesh to exterior of structure using heavy-duty staples.

Plan #30 — Outdoor Owl Enclosure:

Owl enclosure is 24' wide x 15' tall x 60' long. It is similar to plan #29. Upgrade posts to 6x6.

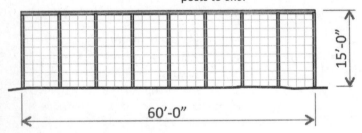

Plan #31 — Outdoor Falcon and Eagle Enclosure:

Falcon and eagle enclosure is 20' wide x 18' tall x 80' long. It is similar to plan #29. Upgrade posts to 8x8.

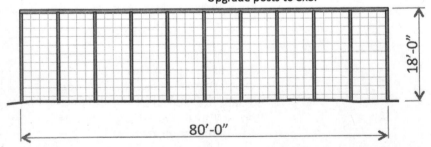

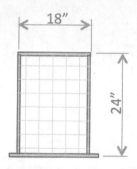

Plan #32 — Small Indoor Tabletop Enclosure:

Small tabletop enclosure suitable for small birds is 18" wide x 18" long x 24" tall.

Step 1: Cut out ½" plywood base, 22" wide x 22" long.

Step 2: Assemble 2x2 frame using wood screws.

Step 3: Install stainless steel wire mesh to wood structure using wood screws and clips.

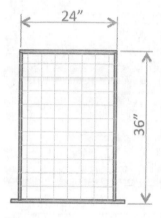

Plan #33 — Medium Indoor Tabletop Enclosure:

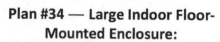

Medium tabletop enclosure suitable for small birds is 24" wide x 24" long x 36" tall.

Step 1: Cut out ½" plywood base, 28" wide x 28" long.

Step 2: Assemble 2x2 frame using wood screws.

Step 3: Install stainless steel wire mesh to wood structure using wood screws and clips.

Plan #34 — Large Indoor Floor-Mounted Enclosure:

Large floor enclosure suitable for parrots and macaws is 36" wide x 36" deep x 60" tall.

Step 1: Cut out ¾" plywood base x 40" x 40".

Step 2: Assemble 2x2 frame using wood screws.

Step 3: Install stainless steel wire mesh to wood structure using wood screws and clips.

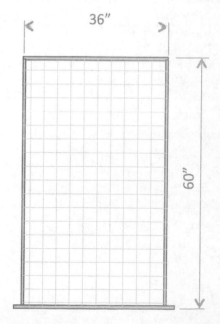

Chapter 5

FISH
AQUARIUMS

Many children's first pets are fish. Everyone has a story or knows someone who won a goldfish at the fair, expecting it to live a couple weeks, and years later the goldfish is still well and thriving. There are all levels of fish ownership, from simply having a few small fish in a bowl of fresh water, to having a large saltwater aquarium of tropical fish that involves careful and detailed maintenance.

Keeping fish as pets dates back thousands of years. It was the Chinese who developed the first goldfish in 960 A.D., and breeding goldfish became a lucrative business in China. It is thought the ancient Chinese had a fascination with bizarre-looking goldfish. These goldfish are known as "fancies" because of their odd physical features. It is thought the Chinese's curiosity with fancies stemmed from their interest in dragons. Later during the 18th and 19th centuries, King Louis XV of France and Prince Potemkin of Russia had goldfish for decorative purposes.

Today, people like to have fish because they are quiet, relatively low-maintenance pets. Some people consider caring for their fish to be a hobby, and many fish owners say that having fish to watch is calming. When you decide you want to own fish, it is important to plan ahead to get their new home ready. You should set up your fish's tank a week in advance so that the water can dechlorinate and bacteria that are beneficial to your fish's health can grow. The ASPCA recommends using a 20-gallon tank because water conditions are easier to maintain with a larger tank. If you are getting tropical fish, you will need saltwater and a heater. The ASPCA suggests beginners start with cold-water, freshwater fish.

Your fish will need a filter in their home. The purpose of the filter is to prevent poisoning from fish waste. In nature, a fish's habitat is hardly ever stagnant, and is never as concentrated as it is in a 20-gallon aquarium. There are several types of filters to choose from, and your tank's size and your preferences should decide which you purchase.

Where you locate your fish's tank is important. Besides making sure the surface you set it on is sturdy, it is important to keep it out of direct sunlight. It should also not be near air vents that can cause great variations in temperature. In nature, the bodies of water fish live in gradually change temperature, unlike the air that comes out of an air conditioning vent. The common goldfish can happily live in water that is between 50 and 70 degrees Fahrenheit, as long as the temperature remains steady. It is wise to use a thermometer to monitor the temperature of the tank; liquid crystal thermometers that attach to the outside of the tank work well and are inexpensive.

To ensure your fish's habitat is as close to a natural habitat as possible, you should provide them with places to hide. In nature, fish hide from their predators in plants, inside hollowed-out rocks, or in driftwood. Your pet store will have many decorative tank accessories that will serve this purpose. Your fish will like hiding in plastic plants, or you could put a "cave" in the tank. Real plants are a nice addition to a tank, but some breeds of fish will nibble on them, which is fine — unless the cost of replacing the plants bothers you. The ASPCA suggests placing an upside-down, clean flower pot with a crack large enough for the fish to enter at the bottom of your tank. You should also put gravel at the bottom of your fish's habitat.

Pet fish also need lighting that is similar to what they would receive in nature. Many aquariums come with lights that are fixed to the bottom of the lid. A florescent light will improve plant growth if you choose to use real plants, and it will also enhance the colors of your fish's scales. To mimic nature, you should have your fish's light on 12 hours a day, followed by having it off for 12 hours.

You may decide that you want several fish. In nature, fish tend to travel in schools, so your pet will appreciate having company. It is essential though to research the breeds of fish you want before introducing new fish to your aquarium. Some breeds tend to be territorial and aggressive and may eat smaller fish. Betta fish are notoriously aggressive, and should be kept alone. It is important you do not overcrowd your aquarium. You should not have more than 1 inch of fish per gallon of water. It is also advisable to add only one to two fish every one to two weeks.

A new home for your fish can vary from a small freshwater indoor tank, to a salt water aquarium, or even an outdoor pond in your backyard. Research the type of fish you want to raise before you make a final decision on a design plan.

Before adding water to your new aquarium, clean the tank thoroughly with a mild soap solution and warm water. Rinse the tank with clean water to ensure any residual soap is flushed away. Gravel used in the bottom of your tank should also be rinsed thoroughly prior to installing. Run clean tap water over the gravel until the water runs clear. Slope the gravel base from the rear of the aquarium to the front. Sloping the gravel can create an illusion that makes the tank appear to be deeper than it actually is. The slope can also aid in cleaning, as large debris will migrate to the front of the tank.

To create a suitable and enjoyable environment for your fish, you can add artificial plants or toys, such as underwater structures that the fish can swim in and around. A hollow floating log will act as a tunnel that fish can swim through. Fish will also explore floating balls on the surface of the water. It may seem like they are playing with the balls while in fact they are just investigating these strange objects. Adding a mirror to the back of your tank can create the illusion of a large aquarium with a greater number of fish. However, do not use a mirror if you own betta fish, as they will become aggravated by what they think is another male fish in the tank and try to attack the mirror.

When you finally fill the tank with water, turn on the filter and run it for a few days prior to purchasing your fish. Running the pump will allow time for the pH level of the tank water to stabilize. Purchase a pH test kit from your local pet store and test the water at least once per month.

Turn on the heater to ensure it works properly. For most tropical fish, the proper temperature is between 75 degrees and 78 degrees. Temperatures below 70 degrees and above 85 degrees can be fatal for most tropical fish. Goldfish are more resilient than tropical fish and can survive in cooler temperatures. A goldfish tank can be kept at room temperature and will not require a heater. Prior to releasing your fish into their new environment, make sure the water temperature of the tank is equal to or close to the water temperature of the bag in which you brought them home from the store. You can even place the bag on the surface of the tank water for a couple hours while the water temperature in the bag stabilizes and matches the temperature of the tank water.

Fish Tank Designs

There are four basic design styles detailed in this book. The first design is an enclosed cabinet with storage for supplies. The second also includes storage below the aquarium; however, it consists of open shelves, similar to a bookcase.

The third design is a double tank configuration. This will allow you to maintain additional species of fish that may not be otherwise compatible with your first breed of fish. For example, you could designate one tank for freshwater fish and use the second tank for saltwater fish.

The final design is a tabletop design. This simple aquarium could be placed on an existing bookshelf, desk, or table.

These design plans can be constructed in five-gallon, ten-gallon, 20-gallon, or 50-gallon configurations. Be sure to decide how many fish you intend to maintain prior to construction so you can determine the proper size aquarium and create a suitable environment for your pet fish.

Selecting the Right Size

There are many factors to consider when determining the size of your fish aquarium. How many fish will you have? What is the size of the largest fish in your aquarium? How often will you need to change the water?

Here are a few basic guidelines to help you determine the size and shape of your aquarium. To give your fish plenty of room to swim back and forth in the tank, the length of the tank should be about ten times the length of the largest adult fish you intend to keep in the aquarium. To provide enough space for your fish to turn around, the width of the aquarium (front to back) should be at least equivalent to the length of the longest fish in the aquarium. The height of the aquarium must be two to three times taller than the height of the largest fish.

The number of fish you can keep in your aquarium varies by type of fish. A general rule for many common types of fish is 1 gallon per inch of fish length. For example if you have two fish and each fish is 3 inches long, you would need a six-gallon tank. However, for certain species of fish that may create more waste, such as an oscar, the guideline may be increased to as much as three to five gallons per inch of fish.

The chart below is a general overview of the recommended number of fish by species per gallon of water.

Species	Average Size	Number of Fish by Aquarium Size			
		5 gal	10 gal	20 gal	50 gal
Angelfish	6"	1	2	3	8
Betta	3"	2	3	6	16
Goldfish	4"	1	2	4	10
Guppy	3"	2	3	6	16
Neon Tetra	1"	4	10	20	50
Oscar	12"	0	0	1	4
Swordtail	4"	1	2	5	12
Tiger Barb	3"	2	3	6	16

Note: Male betta fish should be kept in separate tanks as they will attack each other. If you keep several female fish in the same tank, be sure to provide several hiding places where they can swim to safety.

To determine how many gallons of water your new tank will hold, follow this simple formula:

1 cubic foot = 7.48 gallons

Example 1:

12" H x 12" W x 12" L = 1,728 inches or 1 cubic foot

1 cubic ft x 7.48 gallons = 7.48 gallons

Example 2:

18" H x 18" W x 36" L = 11,664 inches

11,664 inches ÷ 1,728 inches = 6.75 cubic ft

6.75 cubic ft x 7.48 gallons per cubic ft = 50.5 gallons

Fish Aquarium Plans
Plan #35 — 50-Gallon Tank and Cabinet:

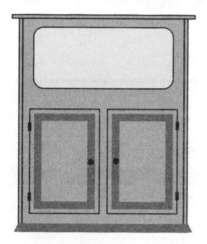

Front View

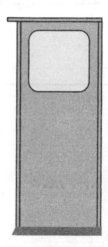

Side View

Bill of Material

Item	Qty	Description	Item	Qty	Description
1	4	2x4 x 1'-3"	14	1	1/2" acrylic x 1'-6" x 3'-0"
2	4	2x4 x 3'-0"	15	2	1/2" acrylic x 1'-5 1/2" x 3'-0"
3	2	1x2 x 2'-9"	16	1	high-strength acrylic structural glue
4	2	1x2 x 11"	17	4	2" hidden hinges
5	1 box	#10 wood screws x 3" long	18	4	finishing trim x 1'-2", cut to suit
6	2	1" plywood x 1'-6" x 3'-0"	19	4	finishing trim x 1'-10", cut to suit
7	1	1" plywood x 1'-8" x 3'-4"	20	2	1/2" plywood x 1'-4" x 2'-0"
8	2	1" plywood x 1'-6" x 4'-0"	21	2	door knob
9	1	1" plywood x 3'-1" x 4'-0"	22	1	piano hinge x 3'-0"
10	1	1" plywood x 1'-6" x 3'-0"	23	1	trim x 3'-3", cut to suit
11	1	1" plywood x 3'-1" x 4'-0"	24	2	trim x 1'-7", cut to suit
12	1 box	#8 wood screws x 1" long	25	1 qt	primer
13	2	1/2" acrylic x 1'-5" x 1'-5 1/2"	26	1 qt	paint

Step 1: Assemble Main Support Frame:

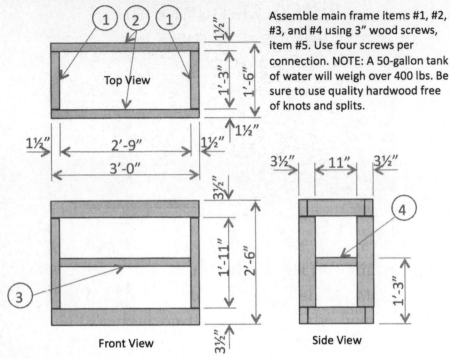

Assemble main frame items #1, #2, #3, and #4 using 3" wood screws, item #5. Use four screws per connection. NOTE: A 50-gallon tank of water will weigh over 400 lbs. Be sure to use quality hardwood free of knots and splits.

Step 2: Cut out Plywood Lid and Bottom:

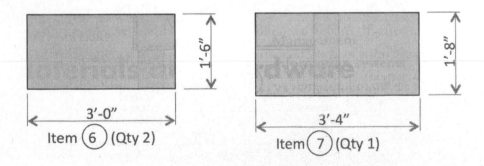

Item (6) (Qty 2)

Item (7) (Qty 1)

Step 3: Cut Out Front and Rear Plywood Walls:

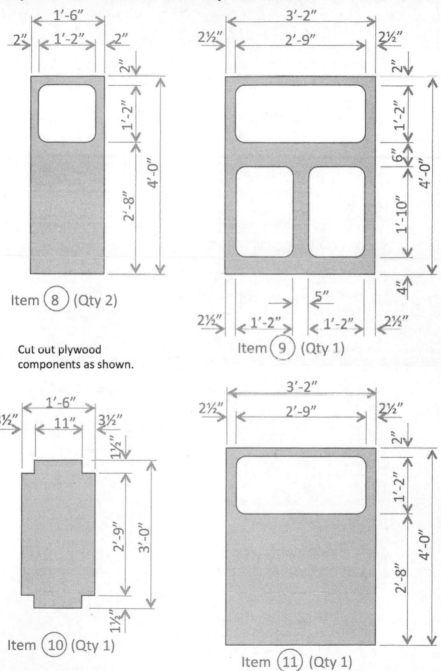

Item ⑧ (Qty 2)

Cut out plywood
components as shown.

Item ⑨ (Qty 1)

Item ⑩ (Qty 1)

Item ⑪ (Qty 1)

Step 4: Attach Top, Bottom, and Shelf to Base:

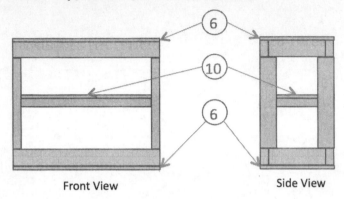

Front View   Side View

Attach panels, items #6 and #10, to base
using 1" wood screws, item #12. Place
screws on 3" spacing.

Step 5: Attach Front, Back, and Sides:

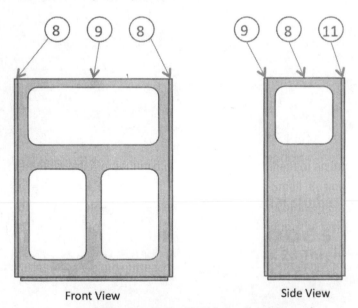

Front View   Side View

Attach panels, items #8, #9, and #11, to
base using 1" wood screws, item #12. Place
screws on 3" spacing.

Step 6: Assemble Acrylic Tank:

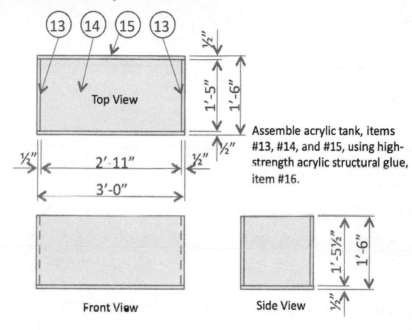

Assemble acrylic tank, items #13, #14, and #15, using high-strength acrylic structural glue, item #16.

Step 7: Install Tank:

Place tank into frame. As an alternative, purchase a ready-made 50-gallon tank. Check dimensions prior to final assembly.

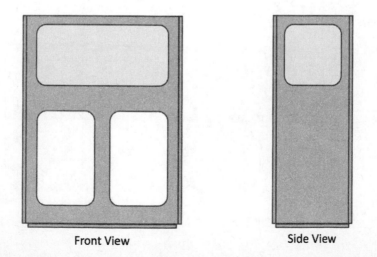

Step 8: Assemble Access Doors (Qty 2):

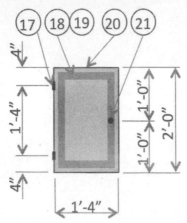

Assemble door as shown using hardware supplied with hinges and knob.

Step 9: Install Trim and Lid; Prime and Paint:

Install doors, attach lower trim, and install hinged lid; prime and paint to suit.

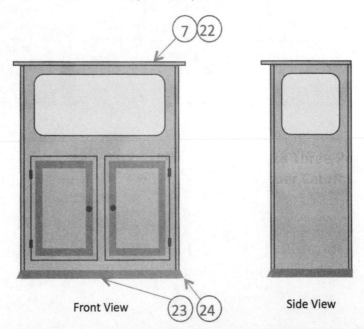

Front View   Side View

Fish Aquarium Plan Variations:

Plan #36 — 50-Gallon Tank With Open Storage:

Similar to plan #35. Remove front doors to create bookshelf-style open access. Add center support for strength. Bill of material is the same as plan #35, less the door material and the addition of the 2x4 front support.

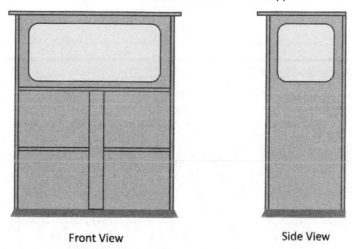

Front View   Side View

Plan #37 — Double 50-Gallon Tanks:

Similar to plan #35. Remove front doors and build second tank at base. Bill of material is the same as plan #35, less the door material, double the quantity of the tank material, and add a hinged lid over the lower tank.

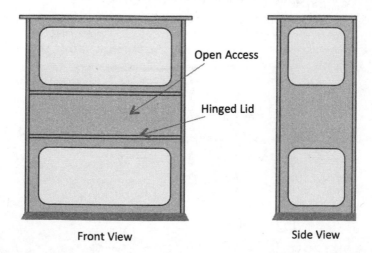

Open Access

Hinged Lid

Front View   Side View

Plan #38 — 50-Gallon Tank Table Top Design:

Similar to plan #35. Remove cabinet below. Bill of material is similar to plan #35. Decrease the height of the panels to dimensions shown and eliminate doors.

Front View   Side View

Plan #39 — 20-Gallon Tank:

Similar to plan #35. Reduce tank size to 14" tall x 14" wide x 24" long. Overall height is 36". Bill of material is the same as plan #35. Reduce the height and width to dimensions shown.

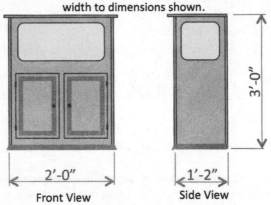

Front View   Side View

Plan #40 — 20-Gallon Tank With Open Storage:

Similar to plan #39. Remove front doors to create bookshelf-style open access. Add center support for strength. Bill of material is the same as plan #35. Reduce the height and width of components to dimensions shown. Add 2x4 front support.

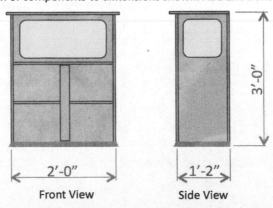

Front View   Side View

Plan #41 — Double 20-Gallon Tanks:

Similar to plan #39. Remove front doors and build second tank at base. Bill of material is the same as plan #35 with dimensions reduced to match plans #39 and #40. Double the quantity of tank components and add a hinged lid.

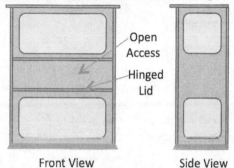

Open Access

Hinged Lid

Front View   Side View

Plan #42 — 20-Gallon Tabletop Design:

Similar to plan #39. Remove cabinet below. Bill of material is the same as plan #35. Reduce the dimensions to match plans #39 and #40 with a tank height of 1'-2":

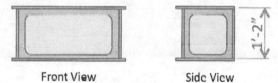

Front View   Side View

Plan #43 — 10-Gallon Tabletop Design:

Similar to plan #42. Tank size is 12" high x 12" wide x 18" long. Bill of material is the same as plan #42 with a tank height of 12".

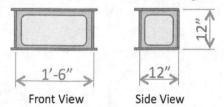

Front View   Side View

Plan #44 — 5-Gallon Tabletop Design:

Similar to plan #42. Tank size is 10" high x 10" wide x 14" long. Bill of material is the same as plan #42 with a tank height of 10".

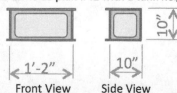

Front View   Side View

Chapter 6

BUILDING A
TURTLE AQUARIUM

Turtle owners range from serious enthusiasts to beach vacationers who buy souvenir-shop turtles on a whim. Turtles first became popular as pets in the mid-20th century when they were "dime-store pets." Turtles were given this name because most pet stores sold these small turtles inexpensively; they came with small, plastic aquariums and fake palm trees. These turtles often did not live long because it was not well known that turtles need specific lighting and water conditions.

In the 1970s, people discovered that turtles carry salmonella, which resulted in the reptile losing popularity as a pet. The Food and Drug Administration (FDA) banned nearly all sales of pet turtles in 1975 when they decided pet stores could not sell turtles with shells less than 4 inches long. Their reasoning behind this was that children could fit the small turtles in their mouths and catch the disease. According to the FDA, nearly 280,000 cases of salmonella poisoning were reported in the United States each year because of these dime-store turtles.

In 2007, the ban against selling small pet turtles was in consideration to be lifted as a result of industry turtle breeders spending about 30 years cleaning up and refining their practices to stop the spread of salmonella. Concordia Turtle Farm, a leader in the turtle industry, saw the percentage of salmonella-carrying turtles decrease from 30 percent to less than 1 percent because of the cleaning system it developed.

To the dismay of pet store owners, the ban on small, baby turtles for pets was not lifted. Larger adult turtles over 4 inches are legal because small children are less likely to cuddle them — and put them in their mouths. The FDA rules state it is legal to sell small turtles for educational, scientific, and exhibit purposes — and many sellers find a loophole by stating they sell their turtles for these purposes. Despite FDA restrictions, small turtles continue to be sold illegally, largely at souvenir shops where store owners do a less-than-satisfactory job of informing tourists how to care for turtles or explain health issues associated with them.

If you find yourself with a pet turtle — large or small — it is important to equip your animal with everything it needs to be healthy and happy. Turtles are much easier to care for than cats, dogs, and other pets, but they are not as vocal about their needs, which often leads them to be neglected. There are both freshwater and sea turtles. Freshwater turtles live mostly on land, and are typically smaller than sea turtles. Sea turtles spend much of their lives underwater, but like freshwater turtles, they must come to the surface to breathe.

Because turtles naturally live in varying environments that consist of both land and water, much time and effort must be put into your pet turtle's home. The correct balance of light, water, moisture, and heat is needed to ensure your turtle is happy. The lighting you provide your turtle is especially important. In the wild, turtles bask in the sun to warm themselves because they are cold-blooded reptiles. Besides warming, sunlight gives turtles essential nutrients like vitamin D 3, which helps them absorb calcium. If your turtle does not receive enough vitamin D-3, it could result in an improperly grown shell, bone disease, and an early death.

In an artificial environment, you need to make sure your turtle is given both UVA and UVB rays. UVA rays are essential to your turtle's appetite, activity level, and breeding. UVB rays, which are responsible for vitamin D-3 production, are especially important. Different breeds of turtles need varying levels of vitamin D-3, so you should research your breed's requirements. If your turtle's breed requires a higher level, you may need to give it calcium and vitamin D supplements, which you can purchase at most pet stores.

The supplements typically come in a powder form that you sprinkle over your turtle's food.

When you are setting up your turtle's home, be sure to buy special UVB bulbs. These should be replaced every ten months because their output weakens with time. Most breeds of turtle thrive under a bulb that emits 3 percent to 5 percent UVB, and you should leave the bulb lit ten to 12 hours a day. You can purchase these bulbs at most pet stores. Research your breed's needs to see if it needs extra hours under the bulb, or if it needs supplements. The UVB bulb should be positioned 12 to 15 inches above your turtle; if you position it more than 18 inches away, your turtle will not receive enough UVB rays. There should be nothing between the bulb and your turtle. A UVB bulb cannot penetrate glass, and mesh will block some of the bulb's UVB rays. If your turtle likes to escape, use mesh with wide gaps so the maximum amount of rays can get through.

When it comes to creating a habitat for your pet turtle, you should research any specifics your turtle's breed needs. There are some basics that both aquatic and semi-aquatic turtles need in their environment. Turtles are often housed in aquariums that are large enough for them to swim, move, and sit under the lamp. A 20-gallon aquarium is needed for two small turtles that are 4 inches. An 8-inch turtle needs a 20-gallon aquarium for itself, and if you have more than one turtle, each adult will need 10 gallons of space.

Because turtles live on both land and water, your aquarium needs to accommodate this. Semi-aquatic turtles need at the most 50 percent water area and the rest land. Aquatic turtles need most of their aquarium to be water and just around 25 percent of it to be land area. The water areas need to be deep enough that your turtle can submerge itself — shell and all — and swim. On average, the water should be around 78 degrees Fahrenheit, but you should research your breed's preferences. To keep your turtle's water clean, use a filter.

Your turtle will come onto land to breathe and warm under the lamp. There are various ways to create land area, including sand, gravel, soil, rocks, and driftwood, or you could build a dock or ramp. You should research your breed's preferences before creating a land area, but no matter what material you use, it is essential to be sure your turtle can get out of the water and onto land easily. Keep the land area around 80 to 85 degrees by using 60-watt basking lamps that have reflectors, and monitor the land's humidity level with a hygrometer. Every breed has different humidity level needs, so research specific requirements for your turtle when creating your turtle's habitat. The size of the land area needs to correlate with how many turtles you have and their size. Also, some breeds can be territorial, which means they need more area to bask under the lamp.

Your turtles will enjoy having plants and logs on their land area to make them feel safer and not as exposed. You can also use aquatic plants in the water because turtles in the wild hide in plants. Some breeds of turtles will eat the vegetation, so it is advisable to have in the aquarium both real and fake plants so you do not have to buy new plants as frequently. However, if your turtle eats the artificial, plastic plants, you should switch to real plants. Be sure to research your breed's plant preferences, and never purchase a plant that could be toxic.

When creating your turtle habitat, there are seven basic design elements that should be included: deep water, a dry landing area, a hiding place, an area of shallow water, lighting, heat, and a filtration system. Your turtle will need a deep tank to swim in. It will also need a dry area or landing to rest on. Place a log or small structure on the landing area so your turtle will have a place to hide.

Female turtles will search for a dry place to lay their eggs. If they cannot find a dry place to bury the eggs, they may carry them too long, potentially causing health risks for both the mother turtle and her young.

Be sure to include an area with shallow water so your turtle can lie in the water but keep its head above the surface to breathe. You should also include lighting and a way to heat the area. Turtles tend to create more waste than fish do, so it is important to purchase an adequate filtering system based on the size and use of your tank. Size the filter for three times the size of your tank. For example, if you have a 20-gallon turtle tank, purchase a 60-gallon filter. An undersized filtration system will affect the health of your turtle.

To create a safe environment for your pet turtle, there are several things that you should avoid. Do not place sharp objects in the tank or on the landing. Flat river rocks would be an excellent choice. Do not use small rocks or pebbles that your turtle could choke on. Do not place toys, rocks, or other structures in such a way that the turtle could become lodged between the glass wall of the aquarium and the toys and accessories. If the turtle gets stuck underwater, it will not be able to surface to breathe.

Artificial plants can be used for decoration. However, if your turtle begins eating the plastic plants, be sure to remove them immediately. Your turtle should diet on leafy vegetables like grape leaves, carrot tops, and lettuce. Feed your turtle carrots, green peas, cantaloupe, grapes, and apples. Foods to avoid include any dairy product such as cheese and yogurt. Never feed your turtle rhubarb or avocado plants, as both these plants are poisonous for turtles. Fresh fruits and vegetables are best, and try to avoid canned vegetables because they tend to have high levels of sodium.

Turtle Tank Designs

When selecting a design for your turtle aquarium, you should plan ahead and select a design style and size that will accommodate the number of turtles you intend to house. You should also consider the space available in your home. Do you have room for a large aquarium, or will you simply add a small aquarium to an existing bookshelf or desk? Will the aquarium be a piece of furniture or just a simple reservoir?

Choose the location for your aquarium in advance. Your tank should be cleaned at least once a week, so consider how close a tank is to a water source and drain source in your home.

It is important to size your tank and landing areas properly for the size and species of the turtles that you intend to maintain. Use these guidelines as minimum requirements for the size of your fully grown turtles.

The capacity of the tank should be at least 5 gallons per inch of turtle. For example, if your turtle is 4 inches long, construct a minimum tank size of 20 gallons. The depth of your aquarium should be greater than the length of your largest turtle. This will allow plenty of space for your turtle to freely turn over in the water. The minimum length of the tank should be three times the length of the turtles and the minimum width must be two times the length of your largest turtle.

The land area should be at least half of the surface area of the water. Again, be sure the design is large enough so the turtles can turn around. Both the length and width of the dry land should be at least one and a half times the length of your turtle. To keep your turtle from escaping, the side walls should be higher than the length of your turtle, or you can cover the top of the aquarium with a lid.

Plan #45 — 36 x 36
with 40-Gallon Tank:

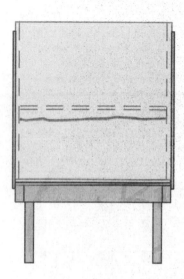
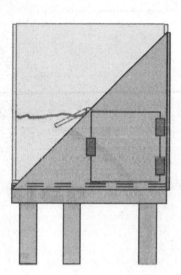

Bill of Material

Item	Qty	Description	Item	Qty	Description
1	3	2x4 x 2'-9"	14	1	1" acrylic x 1'-6" x 3'-0"
2	2	2x4 x 3'-0"	15	1	1" acrylic x 1'-4" x 3'-0"
3	6	2x4 x 1'-6"	16	1	1" acrylic x 1'-7 1/2" x 3'-0"
4	1 box	#10 wood screws x 3" long	17	2	1" acrylic x 2'-11" x 3'-0"
5	1	1" plywood x 3'-0" x 3'-1"	18	1 tube	acrylic cement
6	2	1" plywood x 3'-0" x 3'-0"	19	2	1" plywood x 1'-4" x 1'-4"
7	1 box	#8 wood screws x 1 1/2" long	20	4	2" hinge
8	1	1" plywood x 3'-0" x 3'-0"	21	2	door latch
9	1	1" plywood x 1'-6" x 3'-0"	22	1 qt	primer
10	1	1" plywood x 1'-4" x 3'-0"	23	1 qt	paint
11	1	plastic pan 10" x 10" x 3" deep	24	1	bag of dirt
12	1	1" acrylic x 3'-0" x 3"-0"	25	1	1" acrylic x 4" x 8"
13	1	1" acrylic x 1'-6 1/2" x 3'-0"	26	N/A	accessories and décor to suit

Step 1: Assemble Base Frame:

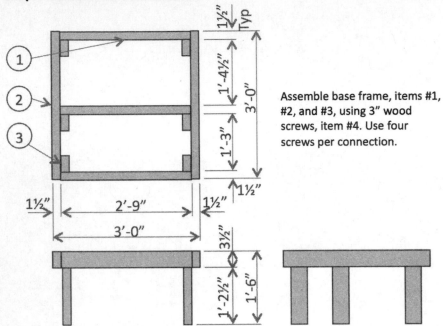

Assemble base frame, items #1, #2, and #3, using 3" wood screws, item #4. Use four screws per connection.

Step 2: Cut out Main Walls:

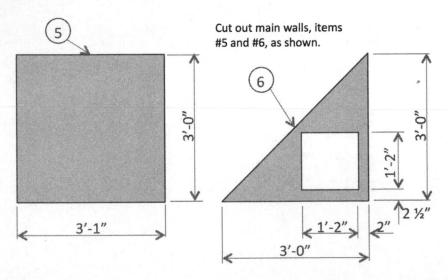

Cut out main walls, items #5 and #6, as shown.

Step 3: Attach Main Walls to Base:

Attach walls, items #5 and #6, to base frame using 1½" wood screws, item #7. Space screws every 3".

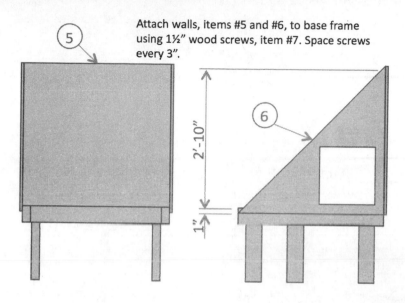

Step 4: Cut Out Inside Supports:

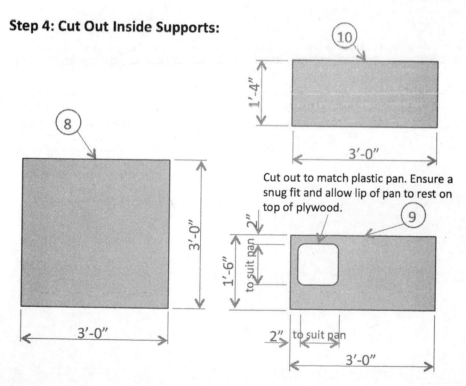

Cut out to match plastic pan. Ensure a snug fit and allow lip of pan to rest on top of plywood.

Step 5: Install Inside Supports:

Install interior supports, items #9 and #10, using 1½" wood screws, item #7. Space screws every 3".

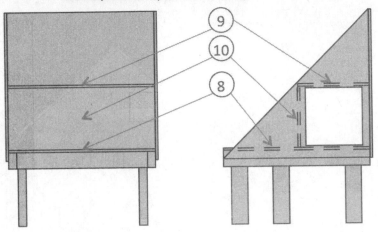

Step 6: Cut Out Acrylic Panels:

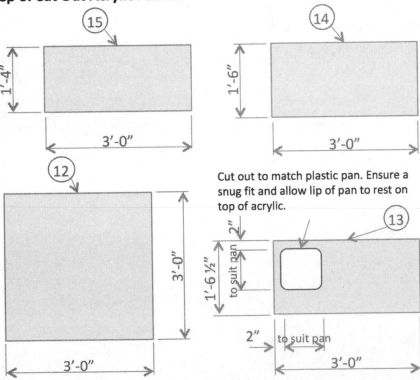

Cut out to match plastic pan. Ensure a snug fit and allow lip of pan to rest on top of acrylic.

Step 7: Cut Out Acrylic Side Panels and Back Panel:

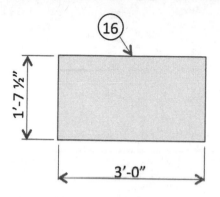

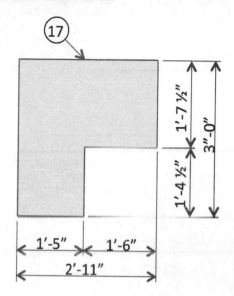

Cut out acrylic side walls, item #17, and back wall, item #16, as shown.

Step 8: Assemble Acrylic Parts:

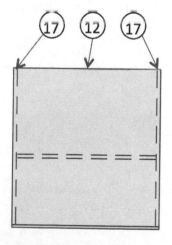

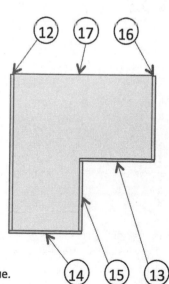

Assemble acrylic tank as shown using high-performance acrylic structural glue. Check fit prior to assembly to ensure finished unit will fit inside wood base structure.

Step 9: Assemble Storage Access Doors (Qty 2):

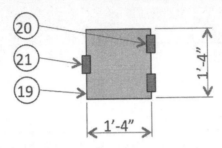

Step 10: Final Assembly:

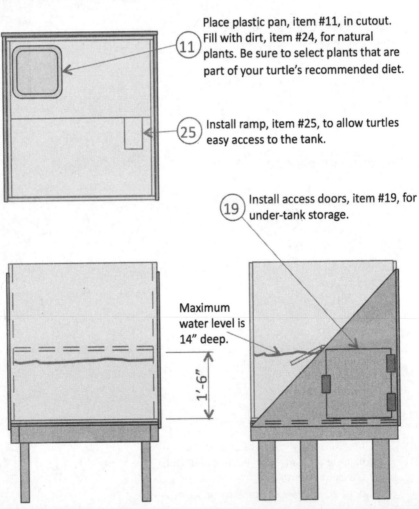

Place plastic pan, item #11, in cutout. Fill with dirt, item #24, for natural plants. Be sure to select plants that are part of your turtle's recommended diet.

Install ramp, item #25, to allow turtles easy access to the tank.

Install access doors, item #19, for under-tank storage.

Maximum water level is 14" deep.

Turtle Aquarium Plan Variations:

Plan #46 — 40-Gallon Tank and Base Only:

Tank is the same size as in plan #45. Remove rear landing and storage. Enclose remaining base and add lower access door. Bill of material is the same as for plan #45. Eliminate side panels and rear panel. Reduce tank side and supports from 36" to 18".

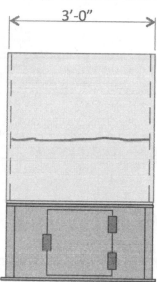

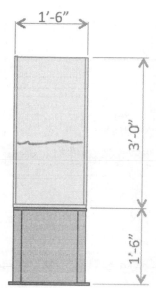

Plan #47 — 40-Gallon Tank with Storage Cabinet:

Similar to plan #45 with additional storage cabinet below.

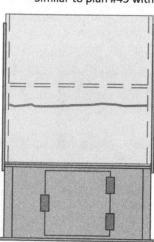

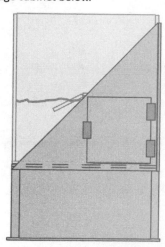

Plan #48 — 40-Gallon Tank, 36 x 48:

Similar to plan #45, but a larger upper landing area. Bill of material is the same as for plan #45. Increase side panels, side panel supports, and tank side from 36" to 48".

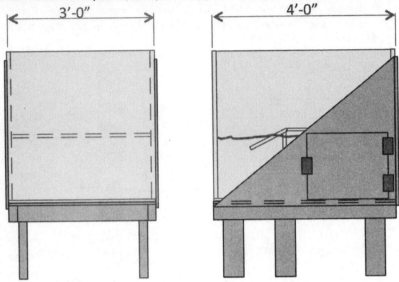

Plan #49 — 20-Gallon Tank, 24 x 24:

Similar to plan #45 with smaller 20-gallon tank. Bill of material is the same as for plan #45. Decrease side panel, supports, and tank components from 36" to 24".

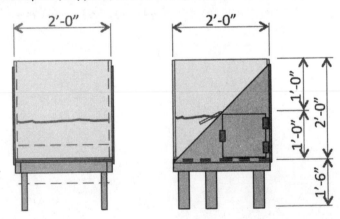

Plan #50 — 20-Gallon Tank and Base Only:

Tank is the same size as in plan #49. Remove rear landing and storage. Enclose remaining base and add lower access door. Bill of material is the same as for plan #46. Reduce side panels, tank side, and support dimensions from 18" to 12".

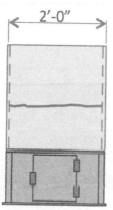
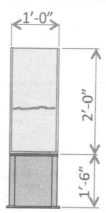

Plan #51 — 20-Gallon Tank and Base Only:

Similar to plan #49, with additional storage cabinet below. Bill of material is the same as for plan #49. Add plywood components to enclose base and add lower access door and hinges.

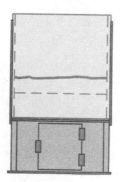
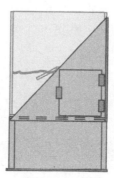

Plan #52 — 20-Gallon Tank, 24 x 36:

Similar to plan #49, with a larger upper landing area. Bill of material is the same as for plan #49. Increase size of landing, side walls, tank side, and supports from 24" to 36".

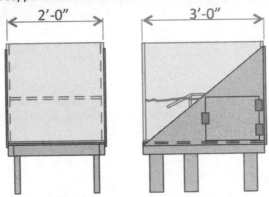

Plan #53 — 65-Gallon Tank, 36 x 60:

Similar to plan #45, with a longer 65-gallon tank. Increase thickness of acrylic to 1¼". Overall length increases from 36" to 60". Bill of material is the same as for plan #45. Increase width of tank components, rear wall, and supports from 36" to 60".

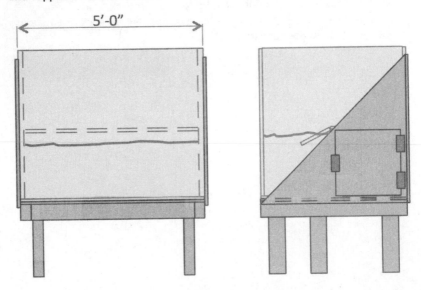

Plan #54 — 65-Gallon Tank and Base Only:

Tank is the same size as in plan #53. Remove rear landing and storage. Enclose remaining base and add lower access door. Bill of material is the same as for plan #53. Eliminate rear panel and side panel, and reduce tank side and support dimensions from 36" to 18".

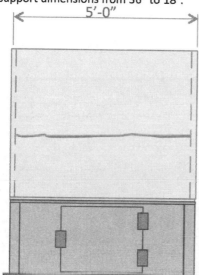
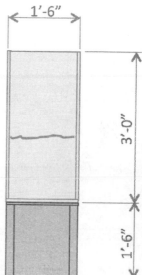

Plan #55 — 65-Gallon Tank with Storage Cabinet:

Similar to plan #53, with additional storage cabinet below. Bill of material is the same as for plan #53, with the addition of lower panels and lower access door.

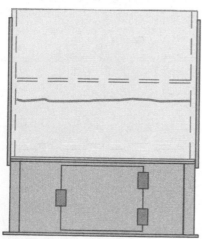
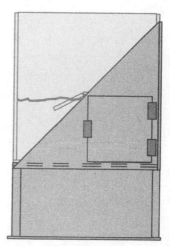

Plan #56 — 65-Gallon Tank, 48 x 60:

Similar to plan #53, with the width increased to 48" to create a
larger landing area. Bill of material is the same as for plan #53.
Increase the dimension of the side panels, tank side, and
supports from 36" to 48".

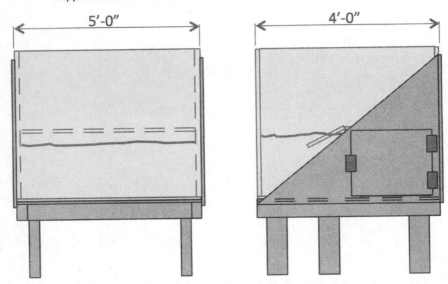

Chapter 7

BUILDING SNAKE PENS

Most people either love or hate snakes, and although having a pet snake might not be for you, it is important to dispel the myth that snakes are slimy, foul creatures. In fact, in some parts of the world, snakes are respected. The ancient Aztecs worshiped Quetzalcoatl, a mythical feathered serpent. In Africa, some cultures worshiped the python, and Australian Aborigines connected the giant rainbow serpent with life and creation. You might wonder why the symbol for modern medicine is a staff with two snakes wrapped around it. Known as "caduceus," this symbol derives from ancient Greek mythology, which says medicine was discovered by Aesculapius, who watched a snake use herbs to bring back to life another snake.

Many people are scared of snakes because of myths that are not true. For example, many people believe that most snakes are poisonous. This is far from the truth, as only about 10 percent of snakes worldwide are venomous. Besides this, there are many folk tales about snakes, including some that say they can suck milk from cows and poison people with just their breath. Stories like this are based on a little truth and a lot of exaggeration.

If you are a snake fan and want to have one or more as pets, there are many things you need to know. Most people keep their snakes in terrariums. Small snake breeds like garter and grass snakes need a ten- to 20-gallon container. Medium-sized snake breeds like king, rat, milk, and gopher need around a 30- to 35-gallon container, and large breeds like boa constrictors and pythons need even larger containers that may need to be custom-built. It is standard for a snake home's length to be three-fourths the snake's length and the home's depth to be one-third the snake's length. The height of your snake's home depends on if it is a terrestrial breed or an arboreal breed. A terrestrial snake stays on the ground, while an arboreal snake has the ability to climb and in the wild would dwell in trees. A terrestrial snake needs its home's height to be three-fourths its total length, and an arboreal snake needs its home to be one times its length, but not longer than 6 to 8 feet. Any larger than this, and your snake will feel uncomfortable and exposed. In the wild, a snake would not sleep out in the open. Instead, it would burrow in the ground or find a small enclosure.

Once in a while when you are driving, you may notice a snake lying in the road. Snakes in the road do not have a death wish, but instead are warming themselves up because, as cold-blooded reptiles, they have no internal way to control their body temperature. Because of this, you are responsible for making sure your snake has the ability to warm up or cool down if it so chooses. You will notice that your snake will move between the cool and warm areas of its enclosure.

It is advised to provide two sources of heat for your snake — primary and secondary. The primary heat source keeps the entire habitat a certain temperature. Incandescent bulbs can be used during the day, and at night you can use nocturnal incandescent bulbs, a ceramic heat emitter, or a heating pad. The secondary source of heat creates warmer areas within your snake's habitat. Use a basking lamp in one area of the terrarium to make a "hot spot," or you can use a heater under the tank. You should give your snake a special place to bask, such as a branch or shelf. You need to research your breed's temperature preferences, and monitor the temperature with a thermometer. A snake that does not eat might be too cold, but if you notice your snake dunking its head in its water, it likely is too hot. The lights in your snake's terrarium provide heat and light. Some breeds of snakes are nocturnal, but just as in the wild, both nocturnal and diurnal breeds need 12 hours of light and 12 hours of darkness a day.

Besides heat and light, humidity is important to your snake. It is essential to research your breed's humidity preferences because in the wild, snakes live in both dry and humid areas. A snake that lives in a tropical rainforest might need a humidity level of 80 percent to 90 percent, while a snake that is from North America might need just 40 percent to 60 percent humidity. Either way, you should monitor humidity with a hygrometer.

After you set up your snake's cage, it is important to include accessories to create an area that is as close to its natural environment as possible. Again, your snake's breed might require specific accessories in its cage, but there are some standard habitat accessories. A bowl of water is needed for more than just drinking. This water should be kept fresh, at room temperature, and it should be in a shallow bowl that cannot be tipped over. Snakes like to soak in their water bowls, and some people provide their snakes with larger containers. If your snake likes to soak, be sure to disinfect the bowl before you refill it each time.

If you see a snake in your yard, you might notice it does not linger or try to approach humans or other predators. For this reason, you should give your snake places it can hide in its habitat so it feels safe and secure. Your snake will also appreciate having shelves and branches to climb so it can bask under the heat lamps. Rocks in a snake's habitat serve a dual purpose. Not only will your snake sit on them to bask for warmth, but it will also rub against them when it is time to shed. This facilitates removal of dead scales — just like humans use pumice stones to remove dead skin.

If your snake's habitat is set up well, it will have a healthy appetite. If your snake is not eating well, its habitat is likely too dark, too cold, or too small. When it comes to feeding, what your snake eats depends on its breed. Some people like to set up a separate place to feed their snakes to keep their permanent enclosure cleaner. If you feed your snake mice, chicks, or rats, a separate feeding enclosure will ensure there are no remnants of the prey's carcass rotting in your snake's home.

Line the bottom of your snake's habitat with sand, carpet, or newspaper for easy cleanup. Unless you want your snake roaming through your home, you need to secure the pen to ensure your snake cannot escape. The access doors on your snake's tank must include a latch, or if your design includes a hinged lid, the access cover must be weighted down.

Your snake pen will require some method of ventilation because all animals require fresh air. Venting the access lid will help control the humidity of the pen, allowing fresh air to enter the enclosure. A series of small holes drilled in the side wall of the enclosure will suffice. Another method is to create a larger hole covered with a fine wire mesh. Ventilate the pen in two opposing locations to create a cross draft.

Choosing a Snake Pen

An unused fish aquarium may be a suitable environment for your snake pen. A more adventuresome person may decide to construct a more elaborate habitat. Be sure to select a design plan that is suitable for the type of snake or species that you intend to house. If you are planning to purchase a snake species that enjoys climbing trees, you should select a design plan with a tall enclosure so you can add a tree or other climbing structure. Terrestrial species do well in a long horizontal enclosure while snakes from the arboreal species prefer a taller enclosure with a climbing structure.

Snakes will require up to 100 square inches of living space for every foot of their length. For most species, a three-quarter square foot enclosure per 1 linear foot of snake will be adequate. As an example, a 1-foot wide by 3-feet long tank would be the minimum size of enclosure that would be adequate for a 4-foot snake. Of course, a larger tank is better if you have the space in your home.

Here is a list of some of the more common pet snakes, including their average adult length, to help you choose a design plan:

- Ball python — 3 feet to 6 feet

- Boa constrictor — 7 feet to 8 feet

- Corn snake — 3 feet to 5 feet

- Garter snake — 2 feet to 3 feet

- Gopher snake — 6 feet to 7 feet

- King snake — 4 feet to 6 feet

- Milk snake — 3 feet to 4 feet

- Rat snake — 3 feet to 4 feet

Plan #57 — 24 x 48 x 60:

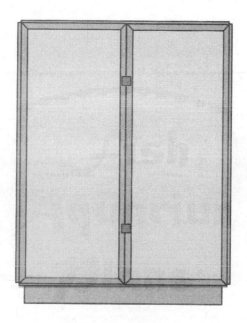

Bill of Material

Item	Qty	Description
1	4	2x2 x 5'-0"
2	8	2x2 x 1'-9"
3	4	2x2 x 3'-9"
4	2	2x4 x 4'-9"
5	1 box	3" deck screws
6	1	1/4" plywood x 4'-0" x 5'-0"
7	2	1/4" plywood x 2'-0" x 4'-0"
8	2	1x4 x 3'-9"
9	2	1/4" Lexan x 2'-0" x 5'-0"
10	2	1x4 x 1'-9"
11	1	1 5/8" deck screws

Item	Qty	Description
12	1 box	1" wood screws
13	4	2x2 x 2'-0"
14	4	2x2 x 5'-0"
15	4	1x2 x 2'-0"
16	4	1x2 x 5'-0"
17	2	1/4" Lexan x 2'-0" x 5'-0"
18	2	piano hinge x 5'-0"
19	2	door latch
20	1 qt	primer
21	1 qt	paint

Step 1: Assemble Main Frame:

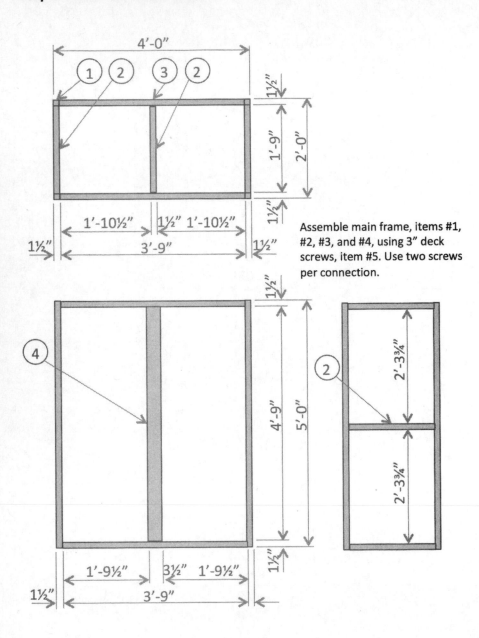

Assemble main frame, items #1, #2, #3, and #4, using 3" deck screws, item #5. Use two screws per connection.

Step 2: Attach Rear Walls, Roof, Floor, and Side Panel to Main Frame:

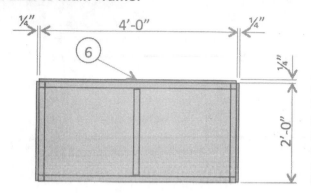

Attach items #8 and #10 to floor, item #7, using 1⅝" deck screws, item #11. Attach rear wall, item #6, roof, item #7, and side panels, item #9, using 1" wood screws, item #12.

Drill ¼" vent holes in back panel equally spaced.

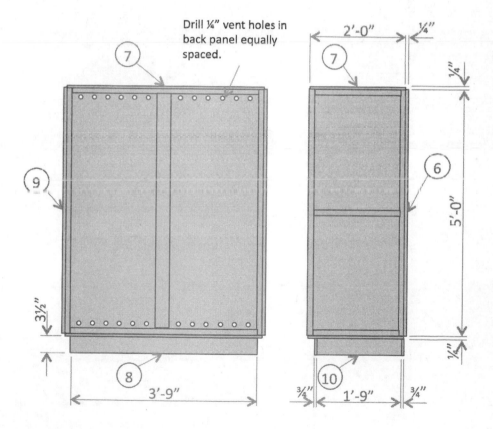

Step 3: Assemble Door Frames:

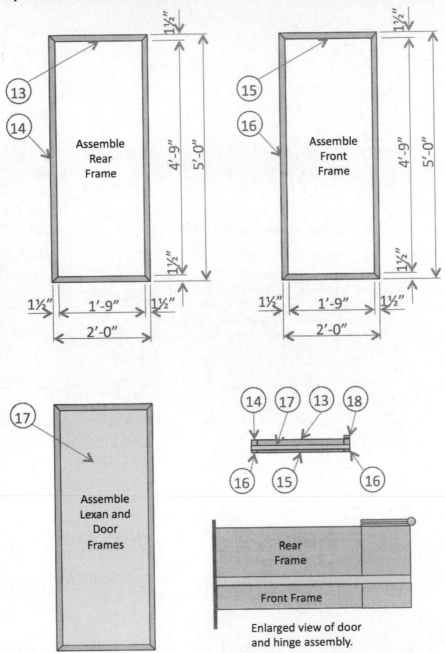

Enlarged view of door
and hinge assembly.

Step 4: Final Assembly:

Install latches, item #19, using hardware provided.

Attach door frame assemblies to main frame using screws provided with piano hinge.

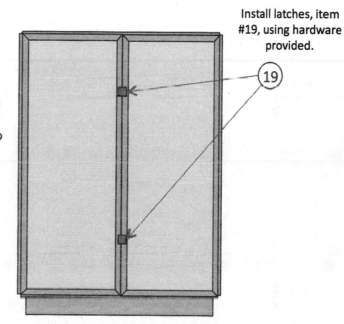

Install climbing tree and heat mat. Prime, paint, and decorate to suit. Be sure to use a pet-safe, non-toxic acrylic paint.

Snake Cage Plan Variations:

Plan #58 — Two-Tier, 24 x 48 x 60:

Similar to plan #57, with a middle divider to create two separate enclosures for different snake species. Construct the divider using ¾" thick plywood that is 2' wide x 4' long.

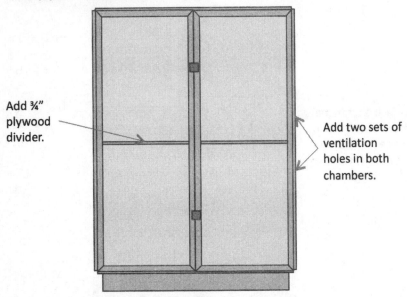

Add ¾" plywood divider.

Add two sets of ventilation holes in both chambers.

Plan #59 — Three-Tier, 24 x 48 x 60:

Similar to plan #57, with two middle dividers to create three separate enclosures for different snake species. Construct the divider using ¾" thick plywood that is 2' wide x 4' long.

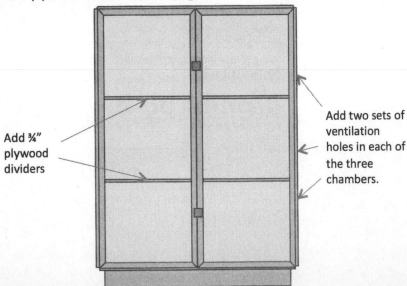

Add ¾" plywood dividers

Add two sets of ventilation holes in each of the three chambers.

Plan #60 — 36 x 36 x 60:

Similar to plan #57. Same overall height, 36" wide x 36" deep.

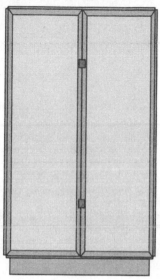

Plan #61 — Coffee Table, 24 x 24 x 48:

Similar to plan #57. Overall dimensions are 24" wide x 24" high x 48" long. For larger snakes, add a latch to the glass top to prevent the snakes from escaping.

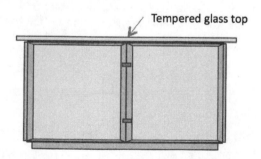

Tempered glass top

Plan #62 — End Table, 24 x 24 x 24:

Similar to plan #61. Overall dimensions are 24" wide x 24" high x 24" long.

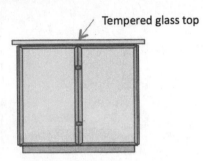

Tempered glass top

Plan #63 — Ten-Gallon Glass Aquarium:

Construct a ten-gallon acrylic aquarium to place on a bookshelf or desk. Overall dimensions are 10" tall x 12" wide x 20" long.

Plan #64 — 20-Gallon Glass Aquarium:

Construct a 20-gallon acrylic aquarium to place on a bookshelf or table. Overall dimensions are 12" tall x 14" wide x 28" long.

Plan #65 — 60 Gallon Glass Aquarium

Construct a 60 gallon acrylic aquarium to place on a bookshelf or table. Overall dimensions are 16" tall x 18" wide x 48" long.

Plan #66 — 36 x 36 x 60:

Similar to plan #57, with multiple dividers. Use as temporary housing for multiple snakes. Insert ventilated, plastic totes to store snakes. Each tote must have its own secured lid.

Add ¾" plywood dividers

Add two sets of ventilation holes in plastic totes.

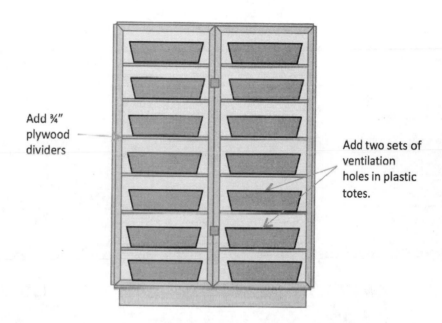

Chapter 8

HOMES FOR
GUINEA PIGS,
HAMSTERS, GERBILS,
AND RABBITS

Guinea pigs, hamsters, gerbils, and rabbits often serve as a child's first pet. Many parents want to start children with one of these animals as a learning experience on life, caring for something, responsibility, and eventually death — what the HSUS refers to as a "starter pet." These furry animals are relatively easy to care for; however, many people dismiss them as extremely low-maintenance, resulting in the animal's needs not being met. Before impulsively purchasing a hamster at the pet store or a rabbit at spring time, it is essential to be equipped with the right knowledge and equipment.

Though small, each of these animals plays in important role in history. For example, the Romans relied on rabbit meat to feed soldiers, and in the Middle Ages rabbits were domesticated and women of the feudal gentry kept them as pets. Later in history, English royalty bred rabbits for sport, and still today they are the most hunted game. Rabbits became popular pets during the Victorian age when the Industrial Revolution made people want to bring aspects of country life into city homes. In the United States, the government urged citizens to raise rabbits for food during World War II, and in the 1980s, Marinell Harriman wrote *House Rabbit Handbook*, popularizing house rabbits.

Humans have had a large part in the development of hamsters. A professor and zoologist from the University of Jerusalem found a litter of golden hamsters in the Syrian Desert. This litter was bred, and today all golden hamsters descend from this litter. Largely bred by humans, hamsters reached the United States in 1938, and are used largely for scientific research because they reproduce rapidly and are mostly disease-free. The domestication of the guinea pig is thought to date back to 5,000 B.C. in Peru, where the rodent served as an essential source of food. The Spanish brought the guinea pig to Europe where it became a popular pet — Queen Elizabeth even had one.

Most pet gerbils descend from the Mongolian gerbil, which missionary, botanist, and zoologist Father Armand

David found in Northern China and 1866 and sent back to the Museum of Natural History in Paris. Humans started breeding the gerbil in 1935, and they reached the United States in 1954. Initially, they were used for scientific research, but soon after became pets.

All these pets are best kept indoors. Some people decide to house rabbits outside, but according to the HSUS, extreme temperatures and predators make the outdoors dangerous for rabbits. Furthermore, the stress of fearing a predator can cause a rabbit to have a fear-induced heart attack. A special, private, and indoor cage for these pets provides them with security and comfort — and equipping their cages with the right amenities will ensure happy and healthy pets.

Essential to your pet's health is the right size cage. This is important for each animal, despite varying sizes. The HSUS warns rabbit owners that the majority of rabbit cages from pet stores are too small; instead, your rabbit's cage should, at minimum, be five times the size of your rabbit. Your rabbit needs enough room to stretch out fully and perch on its hind legs without hitting its head on top of the cage. Building your rabbit's cage will not only save you money, but will also create a better environment than any store-bought habitat.

Also large, guinea pigs need ample room. One guinea pig needs a minimum of 7½ square feet, but more ideal is a 30-inch by 36-inch cage. Having a space that provides enough room is essential to reduce boredom-induced depression, provide room for exercise, make cage cleaning easier, and bring out your pet's natural instincts and personality.

Despite their smaller size, both hamsters and gerbils need adequate room. In the wild, hamsters and gerbils are burrowing animals that create underground tunnels and chambers that serve as protection, shelter, and storage. In particular, hamsters create separate spaces to serve as bathrooms, pantries, and bedrooms of sorts. The more space you provide hamsters and gerbils, the better off your pets will be.

Besides space requirements, it is important to consider and accommodate your pet's natural tendencies. For example, as skittish animals, rabbits need a secluded place to hide — and a cardboard box with an access hole in a rabbit's cage can give it a place to sleep and hide. Rabbits tend to sleep during the day and night, and are most active in the morning and evening. A more outgoing animal, your guinea pig will like being located in the room with the most traffic; however, they do have sensitive ears, so avoid keeping them near the television.

Avoid listening to a squeaky wheel all night by remembering that hamsters are nocturnal animals — and most active at night. Hamsters are also sensitive to temperature. They do best in temperatures between 65 and 75 degrees Fahrenheit. A temperature below 60 F could induce hibernation. Gerbils are known chewers of anything they can get their teeth on. Chewing grinds down their teeth, which never stop growing. Avoid putting plastic in your gerbil's habitat because, if eaten, plastic can cause health problems. Gerbils are also known to be masters of escaping because of their ability to jump well. A mesh covering on the rodent's cage will help prevent any escaping.

These pets all have bedding needs as well. As ground-dwelling, burrowing animals in the wild, rabbits, hamsters, gerbils, and guinea pigs need good bedding. In the wild, these animals are prey animals, which means they are hunted by hungry carnivores. Rabbits and guinea pigs burrow largely for protection and a place to sleep, and they hardly ever use a burrow as a bathroom. The HSUS advises against using cedar and pine wood shavings as bedding because they contain harmful chemicals. Instead, bedding made from paper is a safer option, and there are other varieties of bedding specially designed for small animals available at almost any pet store.

Home Designs

Standard construction materials include galvanized wire mesh, dimensional lumber, and plywood. If wire mesh is used as a floor covering, the maximum opening size should be ½ inch by ½ inch. The roof and side mesh should have maximum openings of 1 inch by 2 inches.

Plan #67 is an elevated cage that is 24 inches wide by 48 inches long and is suitable for one large rabbit.

Plan #68 is also 24 inches wide by 48 inches long, but it is designed to be placed directly on the ground.

Plan #69 is a double-stack combination. Combine one each of plan #67 and plan #68 to create a 4-foot-9-inch-tall stack to house two rabbits.

Plan #70 is a floor-level double stack for two rabbits.

Plan #71 is a floor-level triple stack used to house three rabbits.

Plan #72 is an outdoor pen that is large enough for four small rabbits or two large rabbits.

Plan #73 is a double stack similar to plan #69 — simply add a sloped roof for weather protection.

Selecting the Right Size

A rabbit cage should be four to five times the size of the animal. For small rabbits, a cage that is 24 inches long is adequate. Medium rabbits require a cage that is 36 inches long. For larger rabbits the cage should be 48 inches long to provide ample room for the rabbit to exercise.

Plan #67 — Elevated 24 x 48:

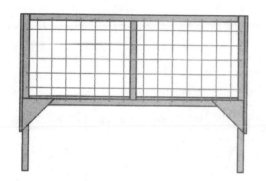 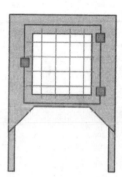

Bill of Material

Item	Qty	Description	Item	Qty	Description
1	4	2x2 x 4'-0"	9	4	1/2" plywood x 7 1/2" x 7 1/2"
2	2	2x2 x 1'-6"	10	1 roll	wire mesh
3	4	2x2 x 3'-0"	11	1 box	heavy-duty staples
4	1 box	3" deck screws	12	1 box	1 5/8" deck screws
5	6	2x2 x 1'-10"	13	2	latch
6	2	1/2" plywood x 2'-1" x 2'-3"	14	4	2" hinges
7	2	1/2" plywood x 1'-6" x 1'-6"	15	1 qt	primer
8	1	1/2" plywood x 2'-1" x 4'-3"	16	1 qt	paint

Step 1: Assemble Main Frame:

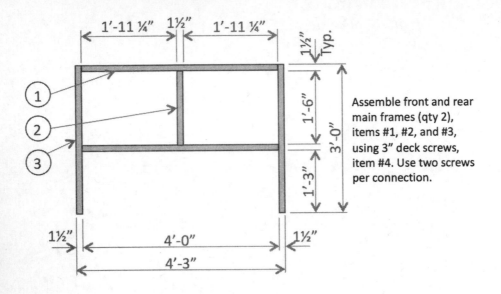

Assemble front and rear main frames (qty 2), items #1, #2, and #3, using 3" deck screws, item #4. Use two screws per connection.

Step 2: Connect Main Frames:

Connect front and rear frames with item #5 using 3" deck screws, item #4. Use two screws per connection.

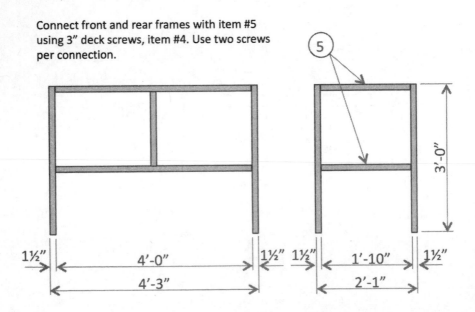

Step 3: Cut Out Doors, Gussets, Floor, and End Panels:

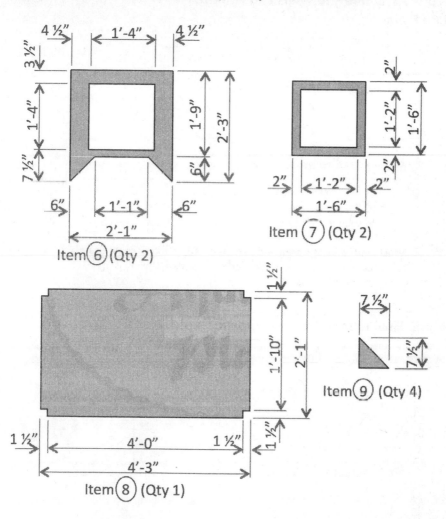

Item ⑥ (Qty 2)

Item ⑦ (Qty 2)

Item ⑨ (Qty 4)

Item ⑧ (Qty 1)

Step 4: Attach Mesh to Access Doors:

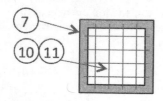

Attach wire mesh, item #10, to end access doors, item #7, using heavy duty staples, item #11.

Step 5: Assemble End Walls and Floor to Main Frame:

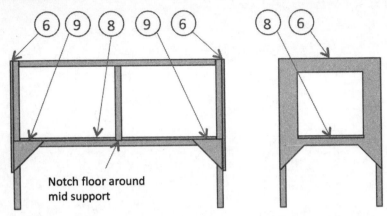

Notch floor around mid support

Attach end walls, item #6, and floor, item #8, to main frame using 1⅝" deck screws, item #12. Place screws every 4". Install item #9 using 1⅝" deck screws.

Step 6: Final Assembly:

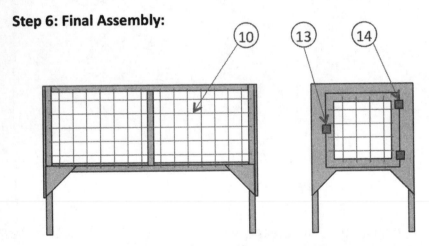

Install wire mesh, item #10, over front, rear, and top opening using heavy-duty staples, item #11. Install access doors using hinges, item #14, and latch, item #13. Prime and paint to suit.

Rabbit Pen Plan Variations:

Plan #68 — Ground Level:

Similar to plan #67. Eliminate legs and place cage on floor or table.

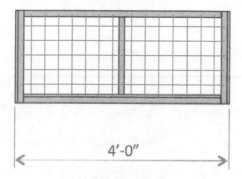

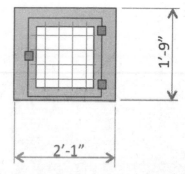

Plan #69 — Double Stack:

Combine plans #67 and #68 to create a stack of two cages for multiple pets.

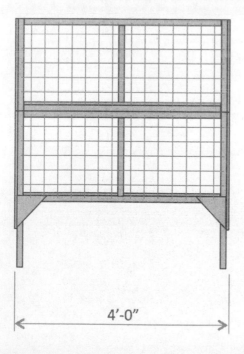

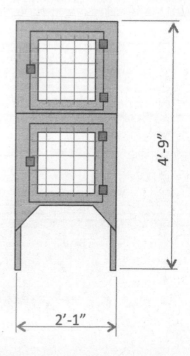

Plan #70 — Ground-Level Stack:

Combine two cages per plan #68 for multiple pets. Place on ground level.

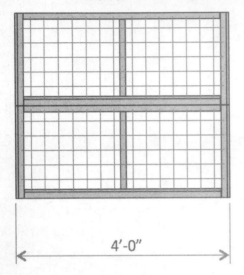

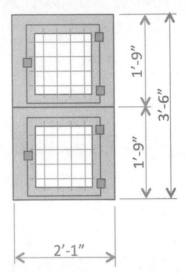

Plan #71 — Triple Stack:

Combine three cages per plan #68 for multiple pets. Place on ground level.

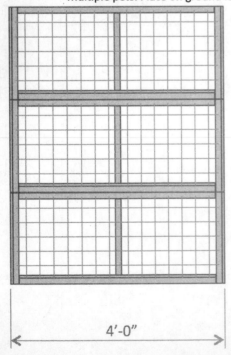

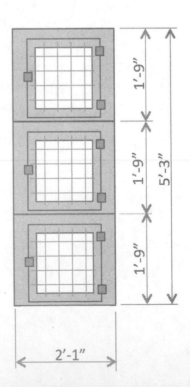

Plan #72 — Outdoor Pen:

Construct a 48" x 48" fence pen.
Assemble corner shelter.

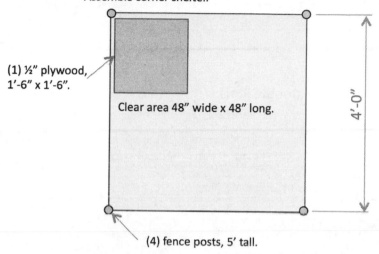

(1) ½" plywood,
1'-6" x 1'-6".

Clear area 48" wide x 48" long.

4'-0"

(4) fence posts, 5' tall.

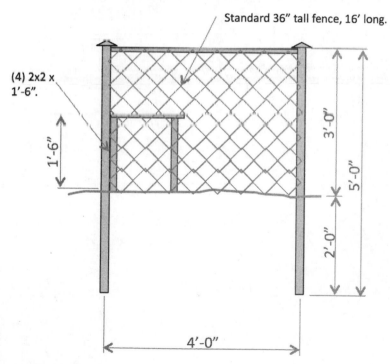

Standard 36" tall fence, 16' long.

(4) 2x2 x
1'-6".

1'-6"

3'-0"

5'-0"

2'-0"

4'-0"

Plan #73 — Outdoor Double Stack:

Similar to plan #69. Add roof for weather protection.

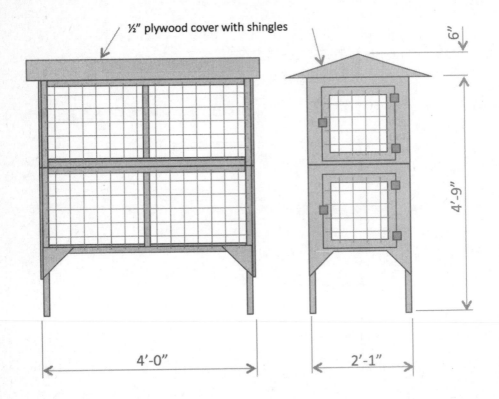

½" plywood cover with shingles

6"

4'-9"

4'-0"

2'-1"

Chapter 9

HOMES FOR CHICKENS, DUCKS, AND GEESE

Having poultry such as chickens, ducks, and geese can be a boon to your family, as these birds provide meat, eggs, and pleasure. Recently, owning chickens has become popular — even some city dwellers are joining in the fun that comes with chicken owning because they appreciate the fresh eggs that chickens produce daily. Waterfowl such as ducks and geese make good pets and enjoy eating weeds that you consider to be a nuisance. Geese learn to guard their territory and can keep intruders off your property.

The modern chicken has its roots in the jungles of Southeast Asia. While Charles Darwin thought all chickens descended from the red jungle fowl, it is now known that the gray jungle fowl also contributed to the gene pool of the domestic chicken. From the breeding of these two wild fowl, we now have hundreds of breeds, types, and strains of chickens. Scientists are not in agreement as to when the chicken was first domesticated. Thailand may hold the distinction of first domesticating the chicken, but many believe there may have been multiple areas in Southeast Asia where wild jungle fowl were captured and kept for egg production. The earliest archeological evidence of domestication has been found in China dating back to 5,400 B.C. Today, chicken meat is the most popular meat in the United States.

The Chinese are credited with first domesticating wild mallards around 1,000 B.C., but ducks were also domesticated around the same time in the Middle East. The ducks were first domesticated for meat and egg use, although later they were used to control pests in rice paddies — a practice still in use today. In the field, the ducks feast on insects, snails, slugs, small reptiles, and waste rice, providing pest control without the use of harsh chemicals.

The goose was domesticated both in eastern Asia and in northern Africa, Europe, and western Asia for its meat, down, and eggs. Archaeological evidence shows that geese were kept in ancient Egypt since 3,000 B.C. The Romans dedicated geese to the goddess Juno. Huge flocks of geese were raised in Western Europe and slowly herded to Rome to supply the city with meat and feathers. As time advanced, large flocks were also raised in southern England, Holland, and Germany, and they were driven to markets in large cities during the fall. Another important product of the goose was the quill, which was used for pens. Geese often act as guards for property. For example, the company that brews Ballantine's Finest Blended Scotch Whisky has been using geese to guard its maturing products since 1959. The guards are nicknamed the Scotch Watch.

Adult chickens have special housing needs that must be met for optimum egg and meat production. The basic requirements for a good chicken coop will protect the birds from the weather and predators, and allow them room to carry out their essential functions: growing, feeding, and egg laying.

You will need to provide your hens with a nesting box that has clean bedding material. Hens need a secluded place to lay their eggs and will lay them in undesirable or hidden spots if not provided with a comfortable nesting box. You can use individual boxes or a community box. As a general rule, be sure your hen can stand in the box comfortably. Community boxes can be any size but need at least two 9-inch by 12-inch openings. The openings should be provided with a flap of material covering two-thirds of the opening to provide seclusion. The nest boxes should be securely fastened to the side of the coop, 2 feet above the floor.

Lighting is also important in the chicken coop. It allows you to be able to observe the conditions inside the coop easily, and it is also essential for egg production because hens need a certain amount of light each day to stimulate egg laying. Under natural lighting conditions, egg laying is maximized during the spring and summer months when there is more daylight. It tends to drop off dramatically in the winter when daylight is short.

Roosts are essential for laying flocks. Chickens have a natural instinct to perch off the ground to avoid predators. This is especially important at night when they are sleeping. Do not use roosts or perches for birds used as broilers, as they may cause breast blisters to form on the chest. A breast blister is a swelling of the tissue in the sternum that occurs in poultry breeds, and a blister will often lead to a downgrade in the meat. Allow 6 to 8 inches of roost space per bird for smaller breeds. Larger breeds should have 8 to 10 inches of roost space. The roost should be constructed of 2-inch by 2-inch wood with rounded edges. Roosts can be arranged horizontally or in a ladder-style arrangement.

Although they are all fowl, ducks and geese have different needs from chickens. Both ducks and geese enjoy water not only for quenching their thirst, but also for cleaning their bodies and exercise. These birds cannot survive without access to bathing water. They will need a deeper trough to dip their bills in to drink than is required for chickens. It is their habit to splash water onto their heads and bills to clean themselves.

If you have a few birds, you can use a hose and a small child-sized pool to provide bathing water. This pool will need to be cleaned and refilled twice each day. The ground surrounding the pool will quickly become muddy with the ducks hopping in and out of the pool. You can combat the mud issue by frequently relocating the pool to a new area or placing it on sand or gravel. The birds should have a waterer for drinking. They should not be able to swim in this water but only submerge their head and drink the water. A small pond can also be used for bathing — place rocks on the bank to prevent erosion.

After six weeks of age, most of a goose's diet can consist of forage, provided by a pasture. Geese are good at foraging both bugs and plants. An acre of pasture can support 20 to 40 geese depending on their age and size. The pasture will need to be monitored, as it can quickly become defoliated and heavily soiled by the geese. Grass for goose pasture should be about 4 inches tall. Longer grass can get bound up in the goose's crop — the outcropping of the esophagus — and cause death. Do not allow the geese to graze the entire pasture down to the dirt. It is best to divide your pasture into five or six separate paddocks and switch the geese to a different paddock frequently.

Ducks and young geese are susceptible to predation, especially domestic ducks that have a limited ability to fly. A predator-proof pen should be provided for your birds. They can roam during the day, but at night you should put them in a pen — nighttime is when predators such as dogs, weasels, coyotes, and raccoons are most active. Strong woven wire should be used with squares smaller than half an inch because weasels are able to squeeze through any wire bigger than this size. The top of the pen should also be covered to prevent animals from climbing over the top. Ducks and geese also will need shade if temperatures approach

70 degrees or above. Natural shade from trees or allowing the birds to access to a covered pen will be adequate.

To improve egg production, a chicken coop should include a nesting area. Chickens prefer a small, cozy area to nest in. For standard-sized chickens, the nesting box should be about 12 inches wide by 12 inches long and a minimum of 10 inches high. For larger chickens, increase the size to 14 inches wide by 14 inches long by 14 inches high. The nesting area should include a 1-inch lip at the front to increase the secure feeling the chickens have while in the nesting area and also to contain the eggs so they do not roll out into the main area of the coop. Building your own coop can be a simple weekend project for you and your family. Most basic designs are constructed using standard building materials available from your local hardware store.

There are many design options to consider when selecting a chicken coop plan. Design styles include a simple box structure similar to a dog house, an elevated coop with an enclosed chicken run, or an easy-to-build "A" frame design.

Ground-level box structure with hip roof: For a quick weekend project, choose a simple box structure with a standard hip roof. Build a square base frame using 2x4 pressure-treated lumber or cedar wood that can be placed directly on the ground. Do not use pressure-treated lumber in areas that may come in contact with your animals. If ingested, the chemicals contained in pressure-treated wood can be harmful. Use ½-inch plywood to build the walls and roof. Cut a 10-inch wide by 12-inch high door opening in the front wall. Cover the roof with standard roofing shingles to provide weather protection. Add a nesting area to the back of the coop with a hinged access cover for easy cleaning and egg retrieval.

Elevated box structure with hip roof: To improve access to the nesting area, elevate a simple box design using standard 2x4 lumber. Raise the coop 18 inches to 24 inches for easy access. Build a wooden ramp that the chickens can use to climb into the coop. Add small strips of wood to the ramp to make it easier for the chickens to ascend the ramp.

Elevated coop with chicken run: To protect your chickens from predators, you can add a fenced-in chicken run. Construct a simple frame using standard dimensional lumber. Cover the frame using standard chicken wire available from your local hardware store or farm supply store. Include a hinged door section for access to the fenced-in area.

Simple "A" frame coop: One of the easiest chicken coops to build is a simple "A" frame design. Using standard 2x4 dimensional lumber, construct an "A" frame structure and cover it with chicken wire.

Selecting the Right Size

When choosing a coop design, you must first determine how many chickens you intend to house. The size of the chicken coop will vary depending on the number of chickens you have. Each chicken requires an area of 3 to 4 square feet of living space and 8 to 10 square feet of running space.

If you will be raising five chickens, you need a coop that offers 15 to 20 square feet of living space and a run area of 40 to 50 square feet. Use the following list to help you determine which of this chapter's plans is best for your circumstances:

Plan #74: 3-foot by 4-foot coop with 8-foot run — three chickens

Plan #75: 3-foot by 4-foot coop with 12-foot run — four chickens

Plan #76: 3-foot by 4-foot elevated coop — four chickens

Plan #77: 3-foot by 4-foot ground-level coop — four chickens

Plan #78: 4-foot by 5-foot with 10-foot run — five chickens

Plan #79: 4-foot by 5-foot elevated coop — six chickens

Plan #80: 5-foot by 6-foot with 12-foot run — eight chickens

Plan #81: 5-foot by 6-foot elevated coop — nine chickens

Plan #82: Tiny starter coop — two chickens

Plan #83: 4-foot by 8-foot "A" frame — three chickens

Plan #84: 5-foot by 10-foot "A" frame — four chickens

Plan #85: 6-foot by 12-foot "A" frame — five chickens

Plan #74 — 36" x 48" with 8' Run

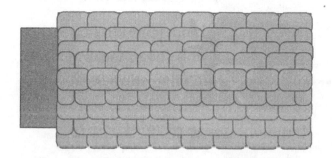

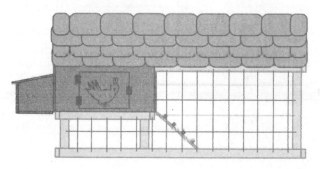

Bill of Material

Item	Qty	Description	Item	Qty	Description
1	6	2x4 x 3'-0"	16	1	1/2" plywood x 1'-2" x 1'-8"
2	6	2x4 x 4'-0"	17	4	1/2" plywood x 1'-2" x 1'-2"
3	4	2x4 x 8'-0"	18	1	1/2" plywood x 1'-0" x 2'-0"
4	2	2x4 x 2'-9"	19	2	1/2" plywood x 2'-7" x 8'-0"
5	1	2x4 x 3'-9"	20	4	2" hinges
6	1 box	3" deck screws	21	1	1 latch
7	1	1/2" plywood x 3'-0" x 4'-0"	22	6	1x1x8"
8	1 box	1 5/8" deck screws	23	1	2x4 x 2'-0"
9	1	1/2" plywood x 2' 11" x 4' 0"	24	2	2x4 x 4"
10	1	1/2" plywood x 2'-11" x 4'-0"	25	1 bdl	shingles
11	1	1/2" plywood x 1'-7" x 3'-0"	26	1 box	1" roofing nails
12	1	1/2" plywood x 1'-7" x 3'-0"	27	1 roll	wire mesh
13	1	1/2" plywood x 1'-2" x 3'-1"	28	1 box	1" heavy-duty staples
14	1	1/2" plywood x 1'-0" x 3'-1"	29	1 gallon	primer
15	1	1/2" plywood x 1'-4" x 3'-3"	30	1 gallon	paint

Step 1: Assemble Base Frame:

Assemble items #1, #2, #3, #4, and #5 using 3" deck screws, item #6. Use four screws per connection.

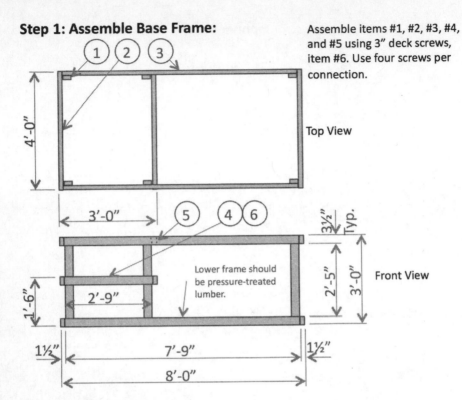

Top View

Lower frame should be pressure-treated lumber.

Front View

Step 2: Cut Out and Install Coop Floor:

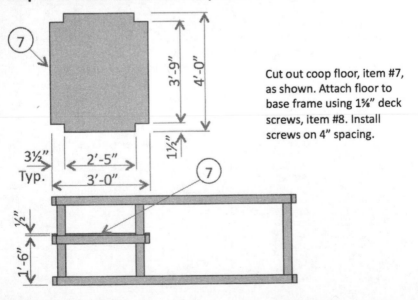

Cut out coop floor, item #7, as shown. Attach floor to base frame using 1⅝" deck screws, item #8. Install screws on 4" spacing.

Step 3: Cut Out Plywood Components:

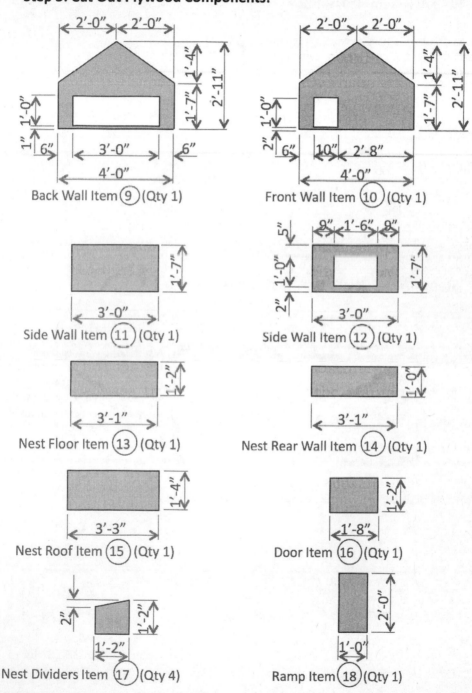

Back Wall Item (9) (Qty 1)

Front Wall Item (10) (Qty 1)

Side Wall Item (11) (Qty 1)

Side Wall Item (12) (Qty 1)

Nest Floor Item (13) (Qty 1)

Nest Rear Wall Item (14) (Qty 1)

Nest Roof Item (15) (Qty 1)

Door Item (16) (Qty 1)

Nest Dividers Item (17) (Qty 4)

Ramp Item (18) (Qty 1)

Step 4: Assemble Nest Box:

Assemble nest box, items #13, #14, and #17, using 1⅝" deck screws, item #8. Install screws on 4" spacing. Space dividers, item #17, equally.

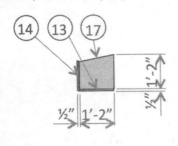

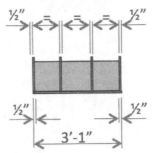

Step 5: Attach Walls and Nest to Base Frame:

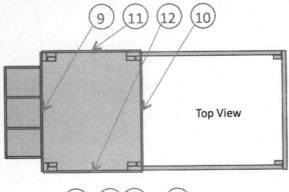

Attach coop walls to base frame using 1⅝" deck screws, item #8. Attach nest box, items #13, #14, and #17, to rear wall using 1⅝" deck screws, item #8. Install screws on 4" spacing.

NOTE: Nest floor should be 2" lower than coop floor.

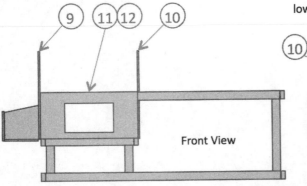

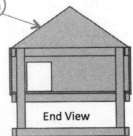

Step 6: Install Roof, Door, Nest Lid, and Ramp:

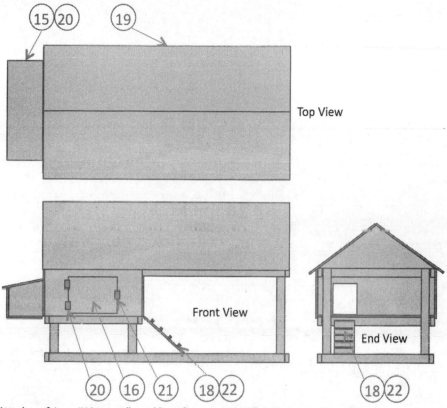

Top View

Front View

End View

-Attach roof, item #19, to walls and base frame using 1⅝" deck screws, item #8.
-Attach nest box lid, item #15, to rear wall using two hinges, item #20.
-Install access door, item #16, complete with latch, item #21, and hinges, item #20.
-Attach ramp, item #18, to base frame using 1⅝" deck screws, item #8.
-Attach cleats, item #22, to ramp, equally spaced, using 1⅝" deck screws, item #8.

Step 7: Assemble Perch:

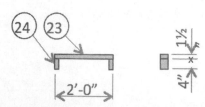

Assemble perch, items #23 and #24, using 3" deck screws, item #6. Use four screws per connection. Attach perch to coop floor using 3" deck screws, item #6. Place it 6" from front wall of coop and 3" from side wall, parallel to front wall.

Step 8: Install Roofing Shingles and Wire Mesh; Paint to Suit:

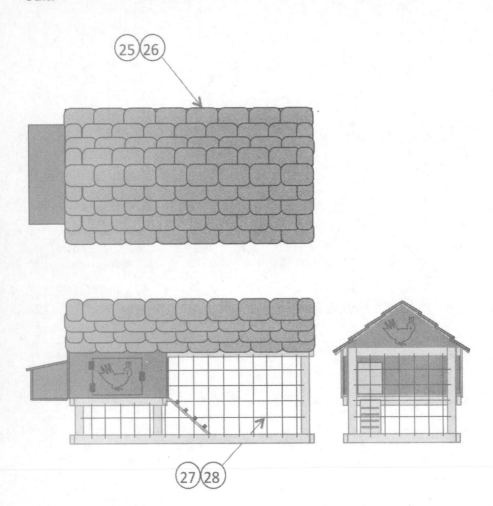

-Install roofing shingles, item #25, to plywood roof using roofing nails, item #26. Follow manufacturer's instructions.

-Attach wire mesh to exterior of base frame using 1" heavy-duty staples.

-Prime and paint to suit.

Chicken Coop Plan Variations:

Plan #75 — 12' Run:

Similar to plan #74 with a 12' chicken run.

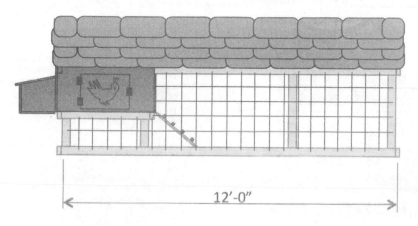

Plan #76 — 36" x 48" Coop:

Similar to plan #74. Eliminate chicken run.

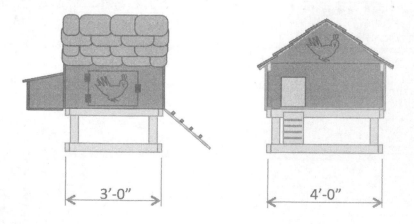

Plan #77 — 36" x 48" Coop, Ground Level:

Similar to plan #76. Shorten legs, item #1, from 3' to 1'-6" and place chicken coop on ground level.

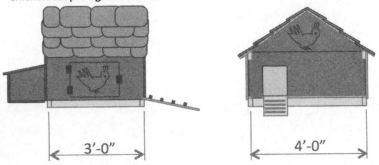

Plan #78 — 48" x 60" Coop:

Similar to plan #74. Increase coop size to 48" wide x 60" long. Chicken run is 10' long.

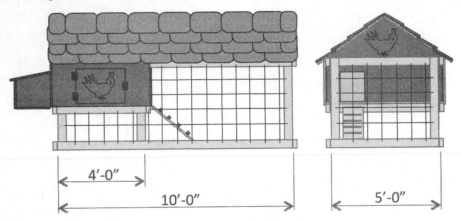

Plan #79 — 48" x 60" Coop:

Similar to plan #78. Eliminate chicken run.

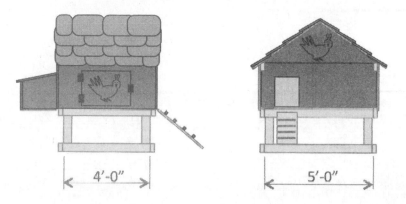

Plan #80 — 60" x 72" Coop:

Similar to plan #74. Increase coop size to 60" wide x 72" long. Chicken run is 12' long.

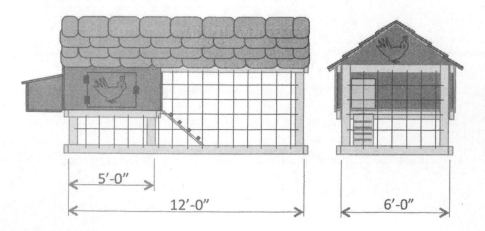

Plan #81 — 60" x 72" Coop:

Similar to plan #80. Eliminate chicken run.

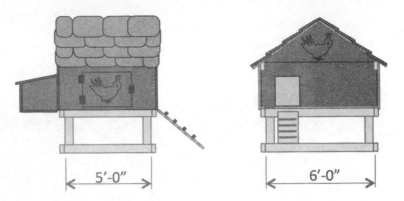

Plan #82 — Tiny Starter Coop:

Similar to plan #74. Eliminate chicken run and nesting box. Overall size is 2'-6" wide x 3' long.

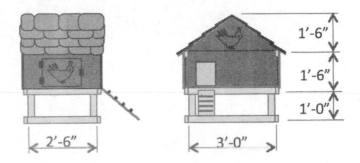

Plan #83 — Small "A" Frame, 48" x 96":

Construct "A" frame structure using 2x4's that are 8' long for the top and the base. Support top using four 2x4's that are 4'-5½" long. Use ½" plywood to build nesting area. Assemble ramp using ½" plywood. Cover with chicken wire.

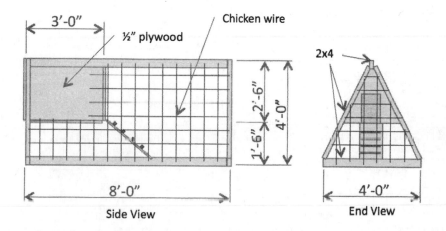

Side View   End View

Plan #84 — Medium "A" Frame, 60" x 120":

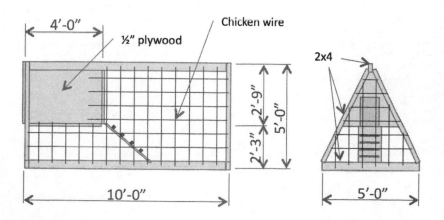

Plan #85 — Large "A" Frame Coop, 72" x 144":

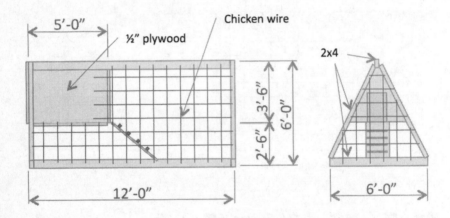

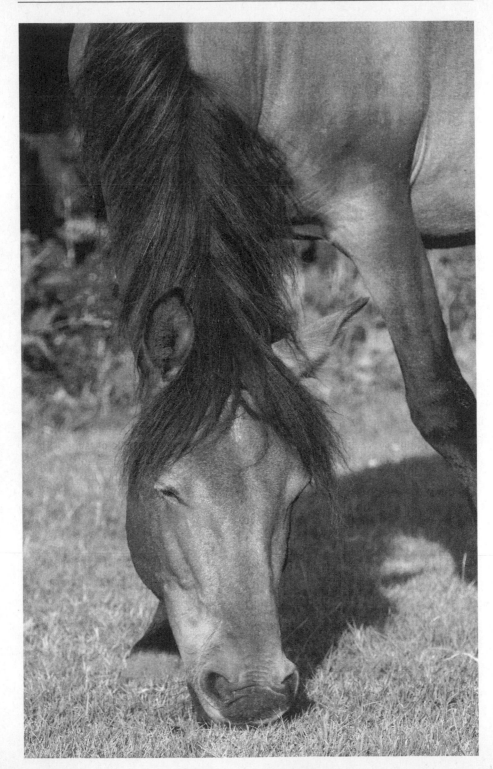

BUILDING A HORSE BARN

Many people dream of owning horses because they are beautiful, make good companions, and provide their owners with many years of fun sport. Before deciding to purchase a horse, it is essential to consider the commitment you are making. Horses are large, powerful creatures that need training. They have expensive needs — the bill for grain and hay alone can be quite steep. If you plan to ride or drive your horses, you will need to purchase all the necessary equipment, which can be a large investment. You will also have to pay for veterinarian bills, farrier costs, dietary supplements, and medications. If you have enough land, one way to save a considerable amount of money is to keep your horse at your house, rather than boarding it at a stable. However, you must be committed to the hobby to do this. If you are unsure, consider first leasing a horse to decide if you have the ability and time to own one.

It is arguable that the horse is the animal that has played the most important role through history. Research suggests that the Botai culture in Kazakhstan domesticated the first horse around 5,500 years ago for riding, food, and milk. This early accomplishment advanced human communication, transportation, and war tactics. Worldwide, the horse has a rich history that has religious, cultural, and economical impacts. In the United States, the horse is an important part of Native American, Colonial, farming, and cultural history. Recent statistics show that there are more than 2 million households in the United States that own horses.

Horses need a diet that mostly consists of grass and hay that is free of dust and mold. Allow your horses to eat as much hay or graze on as much grass as they want. Horses with empty stomachs have a higher risk of developing stomach ulcers, which are common in racehorses. The amount your horse needs to eat will vary, but horses tend to eat between 2 percent and 4 percent of their body weight daily. Your horse will also need access to fresh, unfrozen water throughout the day, even if it only drinks once or twice a day, so ensure you leave a watering trough easily accessible to your horses. Most horses will not need to be fed grains, and some horses develop certain muscle disorders that are linked with increased carbohydrate intake.

These animals were born to run several miles each day, so make sure your horse gets plenty of exercise. Horses can also withstand extreme weather conditions. In the winter, they can withstand temperatures as low as minus 30 degrees Fahrenheit, and in the summer, they can tolerate conditions well over 100 F. However, if the heat index suggests temperatures of over 130 F, use caution when exercising your horse. Your horse's hooves will need to be trimmed every six to eight weeks if it does not receive adequate wear from exercise. Most horses do not need shoes if their hooves are strengthened naturally. Horse teeth grow continuously and wearing unevenly can lead to sharp points and edges that cause difficulty chewing and pain. Your veterinarian will need to check your horse's teeth once or twice a year, and the teeth will need to be floated, or made smoother, as needed.

While horses have survived for centuries in the wild without any structures, you will likely need to offer your horses some form of protection from the elements. On extremely dry and windy days, your horses will need a single wall or windbreak to protect themselves from harsh winds. For rainy, windy, and snowy conditions, horses will need something as simple as a three-sided building with a roof. If the open face is aimed at the most protected direction, which is usually the south, a three-sided shelter should provide adequate protection in most areas. In areas with extreme winter weather, partially closing the fourth side will provide additional protection.

Constructing a horse barn is a major undertaking and requires knowledge of construction and experience using basic woodworking tools and equipment. The key to a successful project is detailed, proactive planning.

One of the most important components of building your barn will be the site you select to build on. Choose a flat, well-drained area and consider the direction of the prevailing winds. While natural winds can improve the ventilation in the barn, you do not want to create a wind tunnel effect, which is the result of the wind hitting the barn head on, increasing the velocity of the wind as it passes through any openings. To avoid this, you may need to orient the barn on a 45-degree angle relevant to the direction of the prevailing winds.

A horse barn is considered a major structure and in most cases will require a building permit from your local code enforcement office. You will be required to provide a detailed drawing of the structure and prepare a dimensioned layout of the construction site. The building department will refer to your plans when it performs its final inspection. It is also important to contact your local utility companies so they can mark the area before you dig any holes. This is a safety measure to ensure that you do not come in contact with any underground wires or pipes. Construction materials you will need to build a barn include standard dimensional lumber. Avoid using cedar because horses tend to eat or gnaw on this type of wood. The most common materials to use are pine and oak lumber because these are the most readily available species and therefore the most cost-effective to use. Be sure to remove any protruding nails or other sharp objects for the safety of your horses.

Roofing options include standard asphalt shingles and metal roofing panels. Asphalt shingles have a life span of 15 to 25 years, while metal roofing panels will most often last for 30 to 50 years. Some metal roofs are rated, or will last, for up to 100 years in certain conditions.

Flooring options include a natural dirt floor, a solid concrete floor, and rubber brick floors. A solid concrete floor is one of the easiest to clean and maintain. A rubber brick floor is also easy to clean and simple to install. Pouring a concrete floor would require heavy equipment and a contractor to prepare the footings and perform the work. You can install a rubber brick floor on your own without the use of any special tools or machinery. The pieces are small and can be easily handled by one person.

Choosing a Design

Standard horse barns vary from a single horse stall to a large barn with multiple stalls. An open run-in can be used for feeding and mid-day weather protection for both individual and multiple horses. An open-sided barn can be built in geographical areas with a fair climate.

For many horse owners, horses are considered an extension of the family, and the owners want to create a comfortable living space for their animals. When selecting a design, you may want to consider an architectural style and materials that are comparable to that of your home or other structures in the immediate area.

Plan #86 features a standard 12-foot by 12-foot stable, an open run-in or feed area, and a covered canopy for storing supplies and farm machinery.

Plan #87 is similar to plan #86, without the canopy. This is a more cost-effective design that still offers weather protection for a single horse.

Plan #88 also offers a single 12-foot by 12-foot stable for a single horse; however, there is a second run-in area that provides a storage location for a tractor or other farm machinery.

Plan #89 features two 12-foot by 12-foot stables for two horses. The open center area is a versatile space that can be used as a run-in area, storage area, or an open garage for your tractor.

Plan #90 is completely open. The three-bay area can be used as a feed area for multiple horses.

Plan #91 is a three-stall stable for housing three horses. Each of the stalls is 12 feet wide by 12 feet long.

Plan #92 is an inexpensive design for a single horse. If your budget is tight, this is a simple plan that can easily be constructed in one weekend and will provide comfortable living accommodations for a single horse.

Plan #93 is a good choice for fair-weather climates. The open design provides comfortable living quarters, fresh air, and overhead weather protection for two horses.

Selecting the Right Size

When designing a horse barn, there are just a few critical dimensions to consider. Each horse stall should be a minimum of 10 feet wide by 10 feet long for smaller horses. A more common stall size is 12 feet wide by 12 feet long.

Enclosed aisles should be a minimum of 12 feet wide to allow larger horses to turn around in the space. The access door should be a minimum of 4 feet wide so horses can easily enter the stalls.

The absolute minimum headroom height is 8 feet for small horses; this is the interior clearance from the ground to the lowest point of the structure. For medium and larger horses, a better choice is 10 feet or 11 feet of clearance.

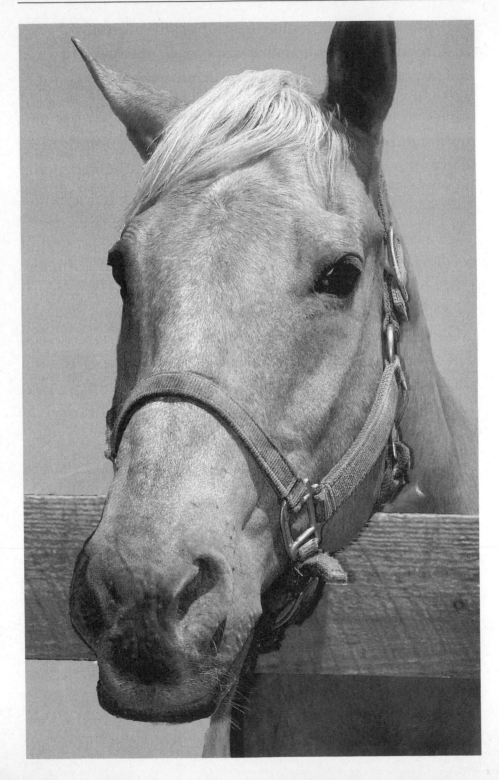

Horse Barn Plans

Plan #86 — Horse Run-In Pole Barn, 20' X 24':

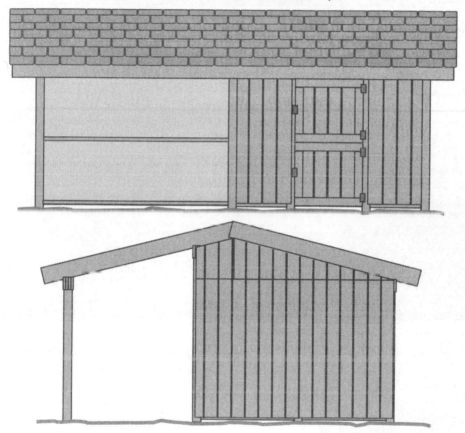

Bill of Material

Item	Qty	Description	Item	Qty	Description
1	11	6x6 post x 12'-0"	13	2 box	10d nails
2	5	6x6 post x 14'-0"	14	2 box	siding nails
3	30 bags	Ready-mix concrete	15	18	1/2" plywood, cut to suit
4	12	2x12 x 26"-0" (spike to suit)	16	6 box	4d nails
5	200	3/8" x 3 1/2" lag screws	17	6 bdl	roofing shingles
6	30	3x3 x 1/4 angle x 6" long	18	6 box	1" roofing nails
7	13	2x6 x 5'-6"	19	8	2x6 x 4'-0", cut to suit for doors
8	42	2x12 x 12'-0"	20	4	3" heavy-duty hinges
9	13	2x6 x 5'-3"	21	2	heavy-duty latch
10	6	2x6 x 3'-11/2"	22	2 gal	primer
11	20	1/2" pine siding, cut to suit	23	2 gal	paint
12	12	2x4 x 3'-93/4"			

Step 1: Layout and Dig Post Holes:

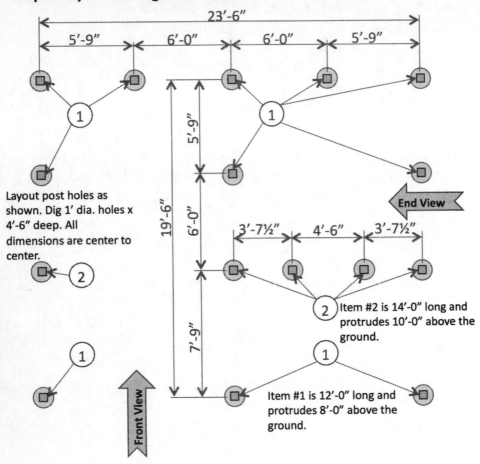

Layout post holes as shown. Dig 1' dia. holes x 4'-6" deep. All dimensions are center to center.

End View

Front View

Item #2 is 14'-0" long and protrudes 10'-0" above the ground.

Item #1 is 12'-0" long and protrudes 8'-0" above the ground.

Step 2: Set Poles in Cement:

Typical hole detail. Prepare hole as shown, insert post, item #1 or #2, and fill with cement. Ensure that all poles are plumb in both directions.

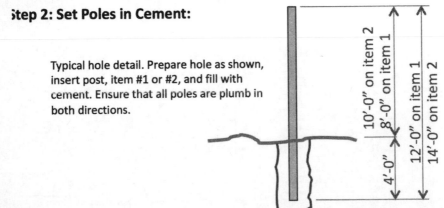

Step 3: Assemble Support Beams:

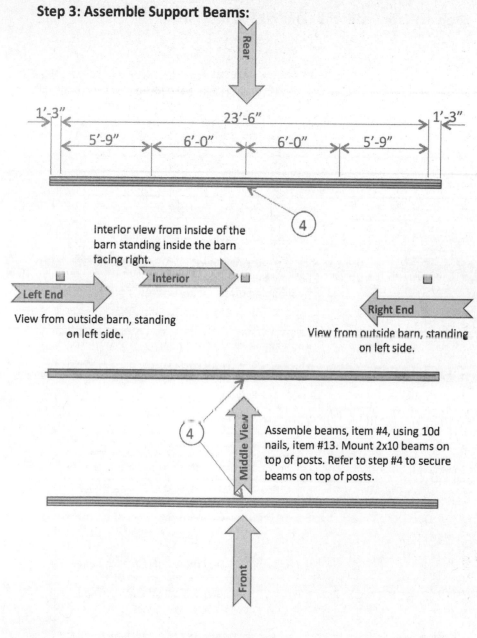

Interior view from inside of the barn standing inside the barn facing right.

View from outside barn, standing on left side.

View from outside barn, standing on left side.

Assemble beams, item #4, using 10d nails, item #13. Mount 2x10 beams on top of posts. Refer to step #4 to secure beams on top of posts.

Plan View

Step 4: Place Support Beams on Top of Posts:

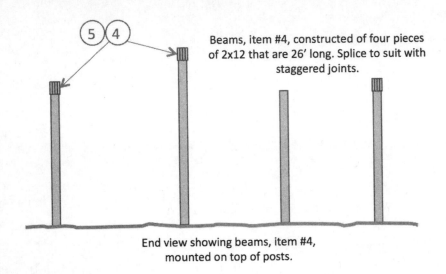

Beams, item #4, constructed of four pieces of 2x12 that are 26' long. Splice to suit with staggered joints.

End view showing beams, item #4,
mounted on top of posts.

Step 5: Fasten Support Beams on Top of Posts:

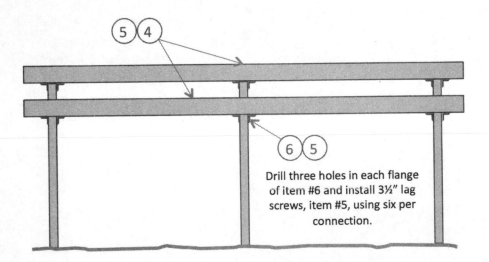

Drill three holes in each flange of item #6 and install 3½" lag screws, item #5, using six per connection.

Front view showing beams, item #4, mounted on top of posts.
NOTE: Some posts removed for clarity.

Step 6: Install Roof Trusses and Wall Supports on End Walls:

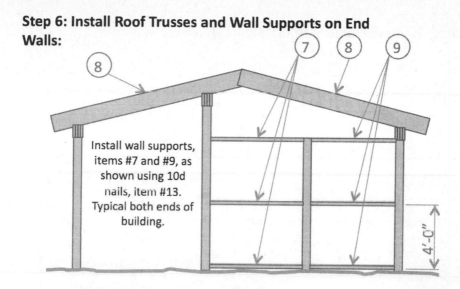

Install wall supports, items #7 and #9, as shown using 10d nails, item #13. Typical both ends of building.

4'-0"

Right end view showing beams, item #4, mounted on top of posts.

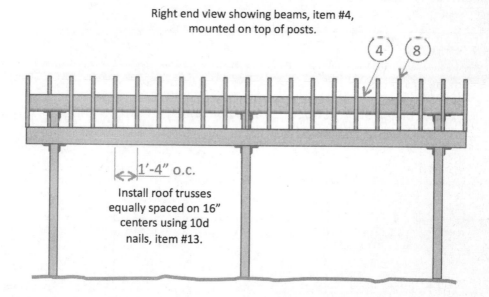

1'-4" o.c.

Install roof trusses equally spaced on 16" centers using 10d nails, item #13.

Front view showing beams, item #4, mounted on top of posts.
NOTE: Far posts removed for clarity.

Step 7: Install Siding Supports on Rear Wall:

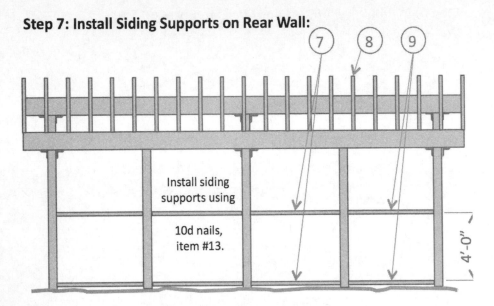

Install siding supports using

10d nails, item #13.

4'-0"

Rear view. NOTE: Far posts removed for clarity.

Step 8: Install Siding Supports on Front Wall:

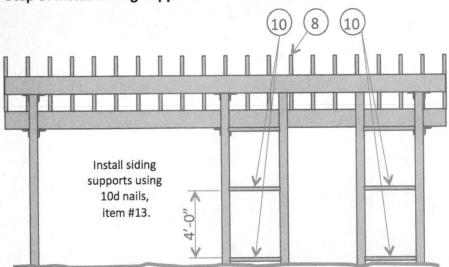

Install siding supports using 10d nails, item #13.

4'-0"

Middle view. NOTE: Far posts removed for clarity.

Step 9: Install Interior Divider Wall Supports:

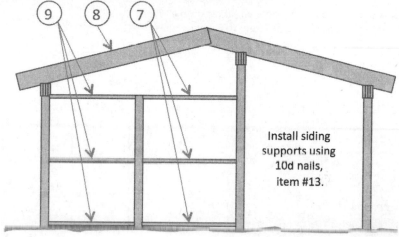

Install siding
supports using
10d nails,
item #13.

Interior view from inside of barn facing
divider wall.

Step 10: Install Interior Divider Wall Lining:

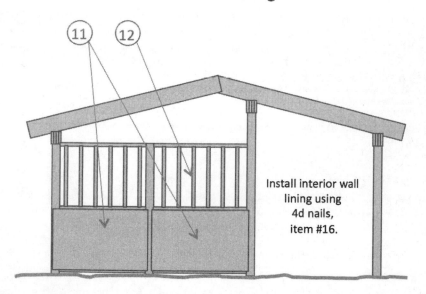

Install interior wall
lining using
4d nails,
item #16.

Left end view from inside of barn.

Step 11: Install Siding and Roof Deck:

Cut roof deck, item #15, and wall siding, item #11, to suit. Install using 4d nails, item #16. Space nails every 6".

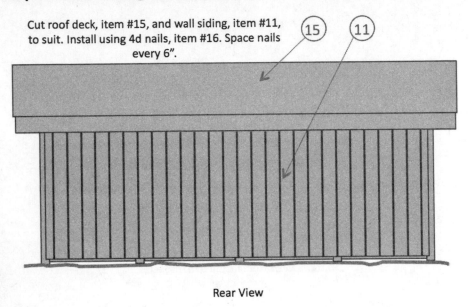

Rear View

Front view. NOTE: Far posts and wall removed for clarity. Front posts and roof removed for clarity.

Step 12: Install Roof Shingles:

Install roof shingles, item #17, using 1" roofing nails, item #18. Follow manufacturer's instructions for nailing pattern.

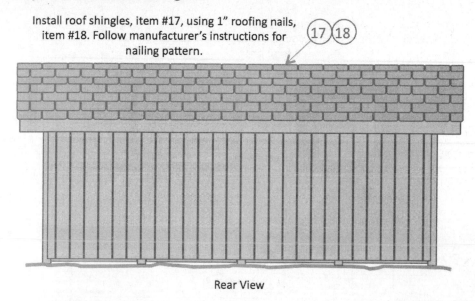

Rear View

This area remains open from the front.

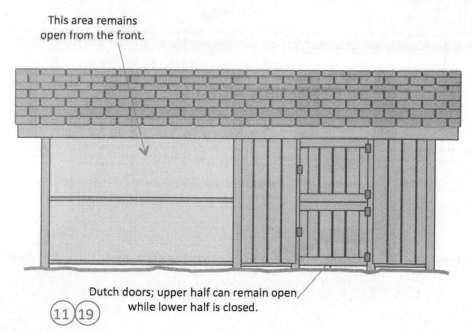

Dutch doors; upper half can remain open while lower half is closed.

Front View

Step 13: End Wall Siding:

Install end wall siding, item #11, using 4d nails, item
#16, on 6" centers. Add windows as desired, and prime
and paint to suit.

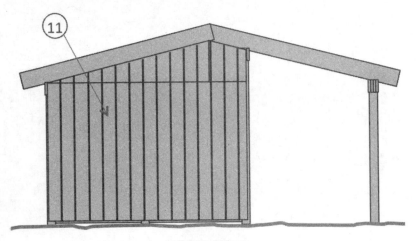

Left End View

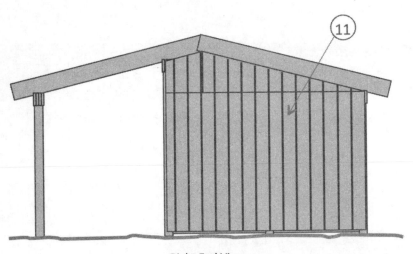

Right End View

Disclaimer: This is a generic plan only. Check local building codes for exact bill of
material requirements. Drawing must be certified by local architect.

Horse Barn Plan Variations:

Plan #87 — 12' x 24', No Canopy:

Similar to plan #86. Eliminate canopy.

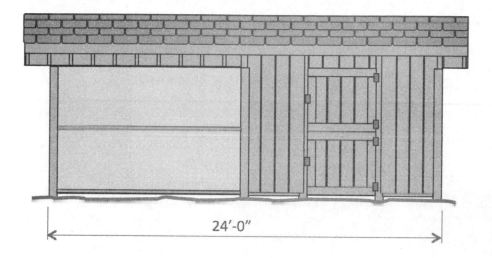

24'-0"

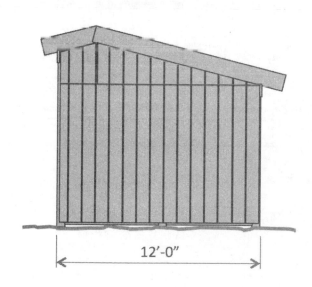

12'-0"

Plan #88 — 20' x 36', No Canopy:

Similar to plan #86, with an additional run in. Standard stable located in the center., and an open run on each end.

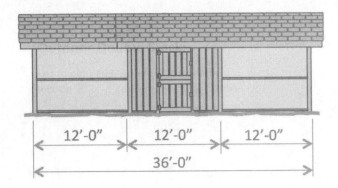

Plan #89 — 20' x 36', No Canopy:

Similar to plan #86, with an additional stable. Open run-in located in middle. Standard enclosed stable on each end.

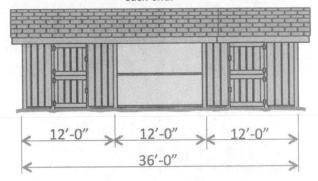

Plan #90 — 20′ x 36′, No Stable:

Similar to plan #88; remove enclosed stable; triple open run-in.

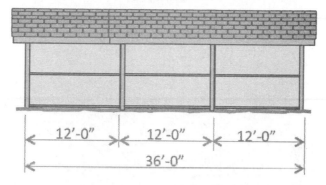

Plan #91 — 20′ x 36′, 3 Stables:

Similar to plan #89, with three standard enclosed stables and no open run-ins.

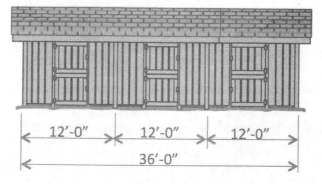

Plan #92 — 12' x 12' Tiny Stable:

Similar to plan #87; stable only — no run-in.

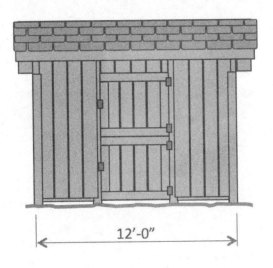

12'-0"

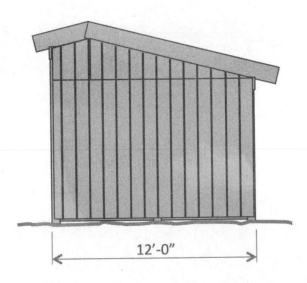

12'-0"

Plan #93 — 20' x 24' Open Stable:

Similar to plan #86, with two open stalls. Remove walls and
replace with 2x4 vertical posts on 6" centers.

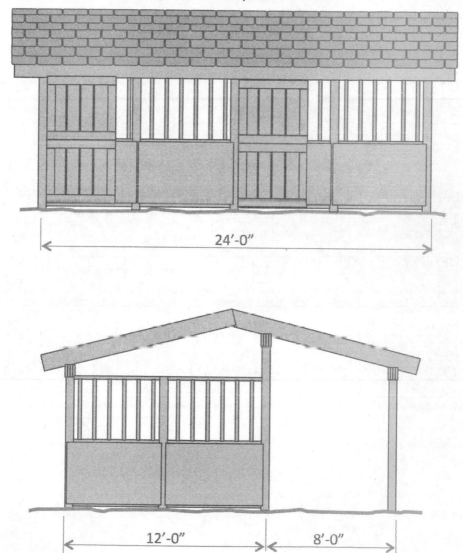

HOMES FOR GOATS AND SHEEP

Goats and sheep were some of the earliest domesticated animals. It is believed goats were domesticated sometime between 6,000 and 7,000 B.C. People enjoy raising goats because the animals have varied personalities and amusing habits. Goats can be affectionate, and during the hot summer months, they enjoy being around their owners and getting the occasional scratch. Most goats enjoy being scratched around their shoulders but some like being scratched between their toes. People can choose to raise goats for either meat or dairy purposes, or just to keep as pets. Some popular goat breeds include alpine, Nubian, and La Mancha.

Many people choose to raise sheep not only as animal companions but also as an additional source of income thanks to their wool. Sheep can survive in the natural environment without a barn or shed. However, it is best to offer them some form of cover to protect against extreme weather situations and for shade during the summer to keep the sheep cool, as they become hot quickly due to their wool coats. Newborn lambs have a hard time retaining body heat so it is essential they receive protection from the cold weather, especially if the conditions are rainy. Be sure to keep their feed dry to avoid mold and other health problems. Many lambs in the United States have their tails docked, or cut close to the edge, to prevent manure buildup that attracts flies. Hooves will need to be inspected and trimmed periodically for foot rot, a fungal infection sheep experience due to standing in mud.

Some people find it difficult to tell the difference between sheep and goats. Some of the ways sheep and goats differ include the following:

- Goats say "maa" while sheep say "baa."

- Goats are more intelligent than sheep.

- Sheep are more likely to overeat than goats are.

- Goats are browsers and sheep are grazers, meaning sheep are better able to eat weeds in your pasture.

- Sheep's milk is higher in fat content than goat's milk. A higher fat content means you are better able to make dairy products such as cheese.

- Sheep are more likely to have worms but are less likely to have external parasites. Goats are more likely to have lice, ringworm, and an internal parasite known as coccidia.

Design your animals' structure with adequate ventilation in mind. When selecting a site for the structure, choose an elevated area with excellent drainage. The easiest floor to maintain would be solid concrete; however, a dirt floor may be warmer and more comfortable for your animals.

Feeder barns require a minimum of 1 foot of feed space per animal.

Plan #94 is 16 feet wide by 24 feet long and can support up to 40 animals.

Plan #95 is 16 feet wide by 36 feet long and can support up to 50 animals.

Plan #96 is 12 feet wide by 16 feet long and can support up to 30 animals.

Plan #97 is a 20-foot wide by 24-foot long center-feed barn that can support up to 50 animals.

Plan #98 is a covered open-sided feed center that can support up to 30 animals.

Plan #99 is a small, covered open-sided feed center that can support up to 20 animals.

Plan #100 is a longer version of the open-sided feed center that can support up to 40 animals.

Plan #101 is a large double-sided feed center that can support up to 100 animals.

Selecting the Right Size

Individual pens within your sheep barn should be a minimum of 20 square feet. A typical pen is 4 feet wide by 5 feet long with 3-foot-high divider rails.

Your barn or feeder should be surrounded by a fence to protect your animals from predators and to ensure they do not wander off. The minimum height of the safety fence should be 4 feet. Proper fencing will have a maximum opening size of 6 inches by 6 inches.

The maximum number of animals per acre ranges from five to ten, depending on the conditions of the land.

Plan #94 — 16' x 24':

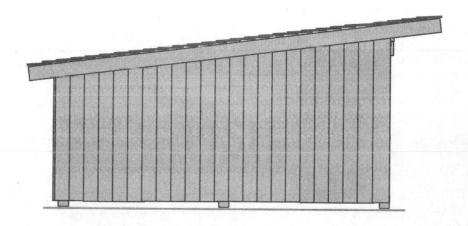

Bill of Material

Item	Qty	Description	Item	Qty	Description
1	4	6x6 x 10'-0"	13	5 bdl	roofing shingles
2	3	6x6 x 11"-0"	14	2 box	1" roofing nails
3	3	6x6 x 12'-0"	15	40	8" siding x 12'-0" long
4	20 bags	ready-mix concrete	16	15	2x4 x 8'-0"
5	8	2x8 x 12'-0"	17	5	1x6 x 8'-0"
6	2 box	10d nails 3" long	18	5	1x4 x 8'-0"
7	6	2x8 x 8'-0"	19	5	1x12 x 8'-0"
8	6	2x4 x 16'-0"	20	11	2x2 x 1'-6"
9	6	2x4 x 8'-0"	21	80	2x2 x 2'-0"
10	19	2x8 x 20'-0"	22	1 gal	primer
11	15	1/2" plywood x 4'-0" x 8'-0"	23	1 gal	paint
12	2 box	6d nails 2" long	24	2 box	1 5/8" deck screws

Step 1: Lay Out and Dig Post Holes:

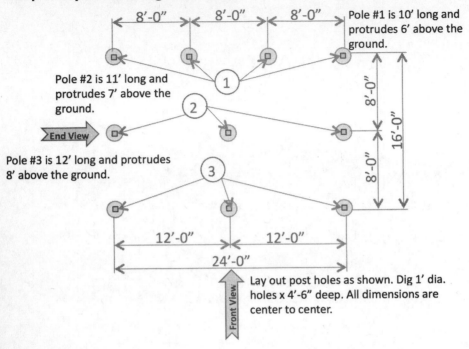

Pole #1 is 10' long and protrudes 6' above the ground.

Pole #2 is 11' long and protrudes 7' above the ground.

Pole #3 is 12' long and protrudes 8' above the ground.

Lay out post holes as shown. Dig 1' dia. holes x 4'-6" deep. All dimensions are center to center.

Step 2: Set Poles in Cement:

Typical hole detail. Prepare hole as shown, insert posts, items #1, #2, or #3, and fill with cement. Ensure that all poles are plumb in both directions.

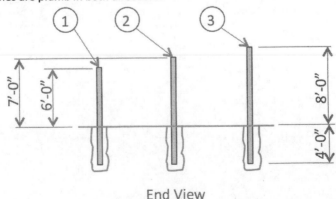

End View

Step 3: Install Roof Supports:

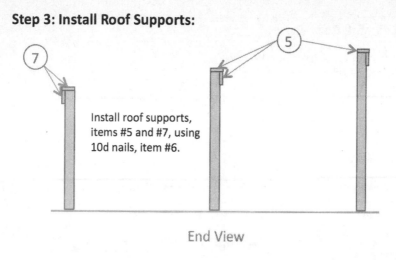

Install roof supports, items #5 and #7, using 10d nails, item #6.

End View

Step 4: Install Wall Supports:

Install wall supports, items #8 and #9, using 10d nails, item #6.

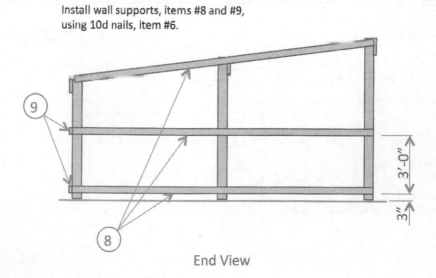

End View

Step 5: Install Roof Truss:

Install roof supports, item #10, using 10d nails, item #6. Space supports on 16" centers.

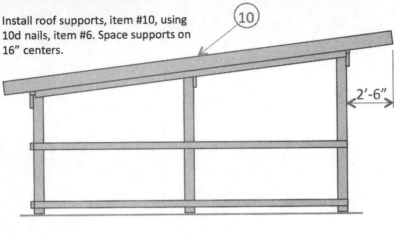

End View

Step 6: Install Roof Deck, Shingles, and Siding:

Install roof deck, item #11, using 6d nails, item #12. Install shingles, item #13, using 1" roofing nails, item #14, per manufacturer's recommended nailing pattern.

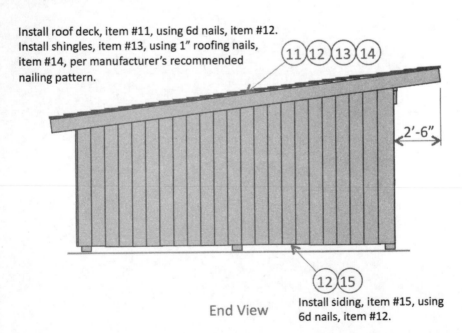

End View

Install siding, item #15, using 6d nails, item #12.

Step 7: Install Feeder:

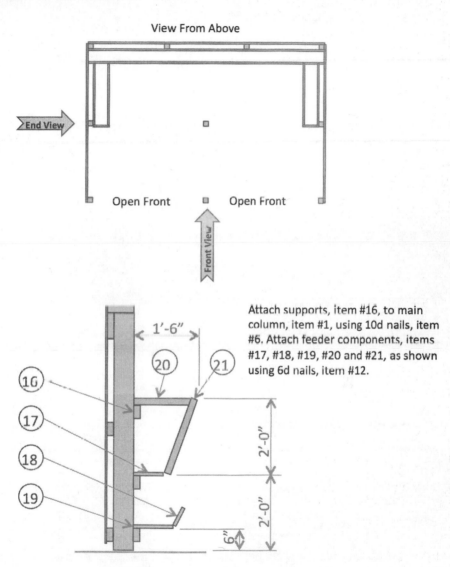

View From Above

End View

Open Front   Open Front

Front View

Attach supports, item #16, to main column, item #1, using 10d nails, item #6. Attach feeder components, items #17, #18, #19, #20 and #21, as shown using 6d nails, item #12.

1'-6"

20   21

16

17

18

19

2'-0"

2'-0"

6"

Enlarged View of Feeder
End View

Disclaimer: This is a generic plan only. Check local building codes for exact bill of material requirements. Drawing must be certified by a local architect.

Sheep Shed and Feeder Plan Variations:

Plan #95 — Three Bays, 16' x 36':

Similar to plan #94, with one additional bay.

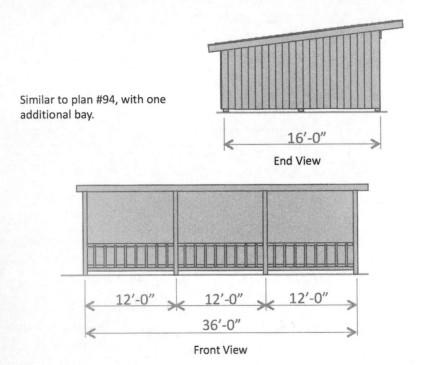

16'-0"

End View

12'-0"   12'-0"   12'-0"

36'-0"

Front View

Plan #96 — Single Bay, 12' x 16':

Similar to plan #94; single bay only.

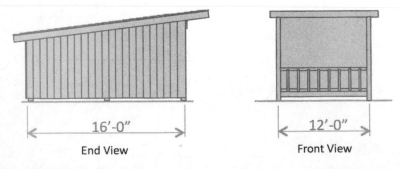

16'-0"

End View

12'-0"

Front View

Plan #97 — 20' x 24' Center Feed Barn:

Install 12' long 8x8 columns as shown. Assemble 2x10 beams using 10d nails that are 3" long. Construct roof trusses using 2x4 material on 16" centers. Add ½" plywood gussets to support roof truss. Cover roof truss with metal roofing material. Install 5' gate in front of raised center feed area. Install two 3' long gates in front of pen area.

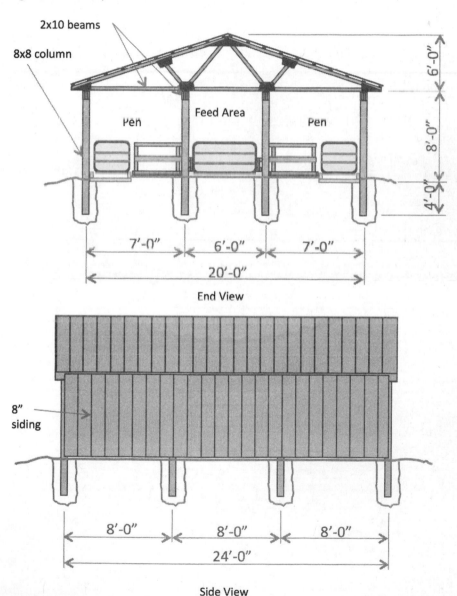

End View

Side View

Plan #98 — 8' x 16' Open Feed Area:

Install 12' long 8x8 columns as shown. Assemble 2x10 beams using 10d nails that are 3" long. Construct roof trusses using 2x4 material on 16" centers. Add ½" plywood gussets to support roof truss. Cover roof truss with metal roofing material. Install 5' gate in front of raised center feed area. Construct open end walls using 2x4 material.

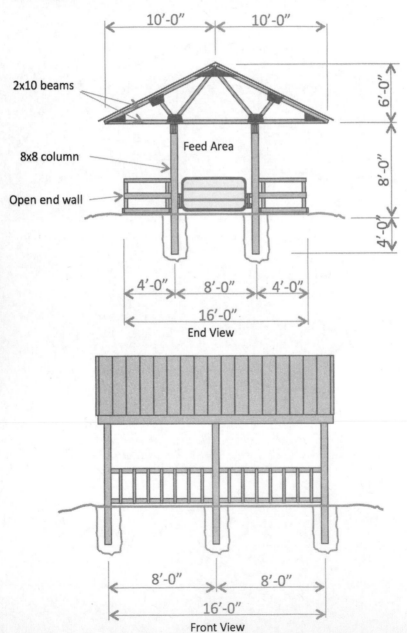

Plan #99 — Open Feeder, 10' Long:

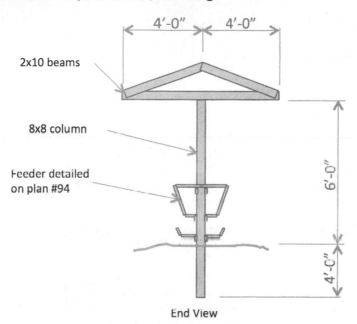

2x10 beams

8x8 column

Feeder detailed
on plan #94

4'-0"   4'-0"

6'-0"

4'-0"

End View

10'-0"

Front View

Plan #100 — Open Feeder, 20' Long:

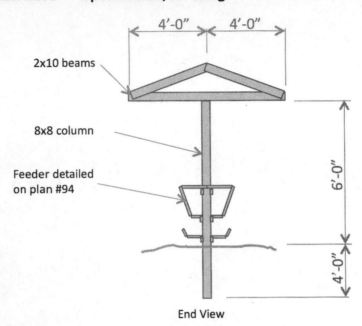

2x10 beams

8x8 column

Feeder detailed
on plan #94

4'-0" 4'-0"

6'-0"

4'-0"

End View

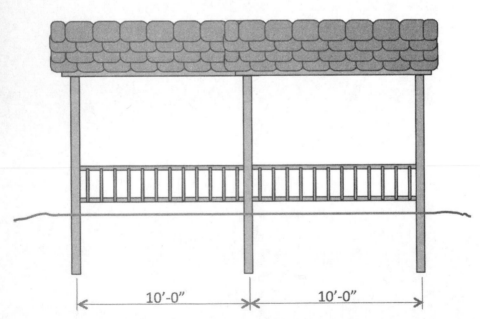

10'-0" 10'-0"

Front View

Plan #101 — Open Feeder, 16' x 16'

Same as plan #94, with the exterior walls eliminated. This version offers double the feeding area by constructing back-to-back feeders. Refer to plan #94 for feeder details; duplicate material on opposite side of column.

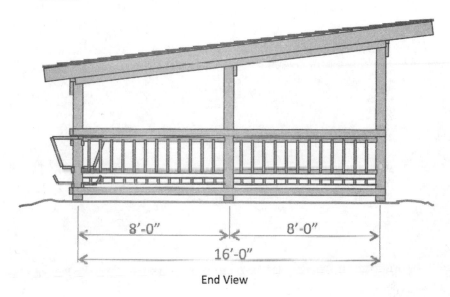

End View

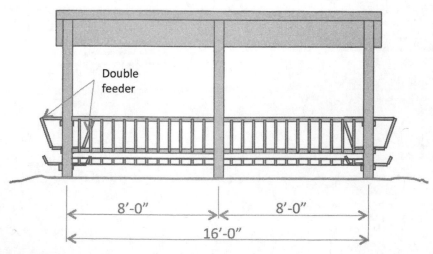

Double feeder

Front View

Conclusion

Building you own animal habitat can be a rewarding experience. With a few basic hand tools and general construction and woodworking skills, you can save a lot of money by building it yourself.

Up-front planning and a little research will ensure you have a successful project. Be sure to check with your local building department to learn if building permits or certified drawings are required. As a courtesy, inform your neighbors of your construction plans prior to getting started.

Have fun and be safe!

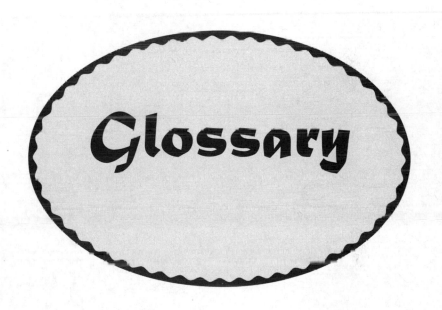

Breast blister: swelling of the tissue in the sternum that happens in poultry breeds; this often leads to a downgrade in the quality of meat

Circular saw: a rotating saw used to cut plywood or framing to size

Clamps: tools used to hold two pieces of plywood or other materials together

Claw hammer: a 13-ounce hammer that is used to assemble 2x4 framework

Den animals: animals that prefer to burrow in enclosed spaces

Dust mask: a covering worn over the face used to prevent the handyperson from inhaling dust particles or other dangerous items

Flight enclosure: a form of housing for birds that features a large fenced-in area for birds of prey that allows the animals plenty of room to fly around while being protected from predators

Floated: the process of making a horse's teeth smoother

Foot rot: a form of foot fungus an animal might experience after standing in manure

Framing level: a tool used to level the main structure of a design

Framing square: a tool used to ensure components fit together and can be assembled more easily

Gambrel roof: a multi-pitch roofline

Hand drill: a handheld power tool used to install screws

Hip roof: a common, single-pitch roof

Hygrometer: a tool used to measure humidity

Lag screw: a form of heavy-duty screw

Metal scribe: a small tool, similar to a screwdriver, with a sharp pointed edge

Nocturnal: mostly active at night

Prey animals: animals that are primarily hunted by carnivorous animals

Roofing hammer: also referred to as a roofing hatchet, this tool has a magnetic end that will hold a nail in place while being used, making it easy for the handyperson to drive in a nail with one hit

Sawhorse: a support used to hold large pieces of wood that are being cut

Sisal rope: a form of rope, similar to twine, used in cat dwellings as an area to scratch

"Aggression Between Cats." The Humane Society of the United States. **www.humanesociety.org/animals/cats/ tips/aggression_between_cats.html**. November 24, 2009. Accessed on June 25, 2010.

Apilado, Crystal. "The Scary Top 7: Readers Reveal What Their Birds Find Scary." BirdChannel.com. **www.bird channel.com/bird-behavior-and-training/bird-behavior -issues/bird-top-scary-things.aspx**. Accessed on June 25, 2010.

"Archaeologists Find Earliest Known Domestic Horses: Harnessed and Milked." Science Daily **www.sciencedaily .com/releases/2009/03/090305141627.htm**. Accessed on October 1, 2010.

"Bird History as Pets." Animal Hospitals USA. **www .animalhospitals-usa.com/birds/bird-as-pet-history .html**. Accessed on June 25, 2010.

"Breeds of Livestock: Goat." Oklahoma State University Board of Regents. **www.ansi.okstate.edu/breeds/goats**. Accessed on October 29, 2010.

"Cat Toys." The American Society for the Prevention of Cruelty to Animals. **www.aspcabehavior.org/articles/ 106/Cat-Toys.aspx**. Accessed on June 25, 2010.

"Choose the Right Birdhouse." National Wildlife Federation. **www.nwf.org/News-and-Magazines/ National-Wildlife/Birds/Archives/2010/Best-Bird -Houses.aspx**. Accessed on June 25, 2010.

"Destructive Scratching." The American Society for the Prevention of Cruelty to Animals. **www.aspcabehavior .org/articles/25/Destructive-Scratching.aspx**. Accessed on June 25, 2010.

"Do You Chain Your Dog?" The Humane Society of the United States. **www.humanesociety.org/issues/ chaining_tethering/tips/do_you_chain_your_dog.html**. October 21, 2009. Accessed on June 25, 2010.

"Dog Care Essentials." The Humane Society of the United States. **www.humanesociety.org/animals/dogs/tips/ dog_care_essentials.html**. Accessed on June 25, 2010.

"Dogs: Positive Reinforcement Training." The Humane Society of the United States. **www.humanesociety .org/animals/dogs/tips/dog_training_positive _reinforcement.html**. October 26, 2009. Accessed on June 25, 2010.

"Every Dog Needs a Den." American Humane Association. **www.americanhumane.org/protecting-animals/ adoption-pet-care/care/dog-dens.html**. Accessed on June 25, 2010.

"Fish Care." American Animal Hospital Association. **www .healthypet.com/PetCare/PetCareArticle.aspx?art _key=c2a2c23e-f33a-4e8b-9965-5590b690925f**. Accessed on August 25, 2010.

"Fish Care." The American Society for the Prevention of Cruelty to Animals. **www.aspca.org/pet-care/small-pet -care/fish-care.html**. Accessed on August 25, 2010.

Forstadt, Michael S. "History of the Guinea Pig." **http:// cavyhistory.tripod.com**. Accessed on October 1, 2010.

"Frequently Asked Questions: Snakes." Doctors Foster and Smith. **www.drsfostersmith.com/pic/article.cfm ?c=6016&articleid=2611&d=160&category=382 #answer_9**. Accessed on August 25, 2010.

Furtman, Michael. *Why Birds Do That*. Willow Creek Press: Minocqua, WI. 2004.

"General Bird Care." The American Society for the Prevention of Cruelty to Animals. **www.aspca.org/pet -care/small-pet-care/bird-care.html**. Accessed on June 25, 2010.

"Gerbil Housing." The Humane Society of the United States. **www.humanesociety.org/animals/gerbils/tips/ gerbil_housing.html**. Accessed on October 1, 2010.

"Goats." Irvine Mesa Charros 4-H Club. **www.goats4h .com/Goats.html**. Accessed on October 29, 2010.

Greelis, Jim. "Pigeons in Military History." **www.pigeon center.org/militarypigeons.html**. Accessed on June 25, 2010.

"Guinea Pig Housing." The Humane Society of the United States. **www.humanesociety.org/animals/guinea_pigs/ tips/guinea_pig_housing.html**. Accessed on October 1, 2010.

"Hamster Housing." The Humane Society of the United States. **www.humanesociety.org/animals/hamsters/ tips/hamster_housing.html**. Accessed on October 1, 2010.

"History of Hamsters." Hamsterhideout.com. **www .hamsterhideout.com/history.html**. Accessed on October 1, 2010.

"Housing for Your Pet Bird." The Humane Society of the United States. **www.humanesociety.org/animals/pet _birds/tips/pet_bird_housing.html**. October 21, 2009. Accessed on June 25, 2010.

Jacmenovic, Michelle. "Baby Turtles and Children: A Dangerous Combination." The Humane Society of

the United States. **www.humanesociety.org/issues/
exotic_pets/facts/baby_turtles_children_090204.html**.
September 2, 2004. Accessed on August 25, 2010.

"Keeping Your Cat Off Countertops and Tables." The
American Society for the Prevention of Cruelty to Animals.
**www.aspcabehavior.org/articles/110/Keeping-Your
-Cat-off-Countertops-and-Tables.aspx**. Accessed on
June 25, 2010.

"Lighting for Turtles and Tortoises: Why UV is Key."
Doctors Foster and Smith. **www.drsfostersmith.com/
pic/article.cfm?aid=827**. Accessed on August 25, 2010.

"Make Your Garden Bird Friendly." Better Homes and
Gardens. **www.bhg.com/gardening/design/nature
-lovers/grow-a-bird-friendly-garden/?page=11**.
Accessed on June 25, 2010.

"Mongolian Gerbil History." The Mongolian Gerbil
Website. **www.petermaas.nl/gerbils/origin.htm**.
Accessed on October, 1, 2010.

"Rabbit History." Rabit World View. **www.rabbitworldview
.com/rabbithistory.php**. Accessed on October 1, 2010.

"Rabbit Housing." The Humane Society of the United
States. **www.humanesociety.org/animals/rabbits/tips/
rabbit_housing.html**. Accessed on October 1, 2010.

Rovner, Julie. "Bill Seeks to Lift Ban on Baby Pet Turtles."
National Public Radio. **www.npr.org/templates/story/**

story.php?storyId=10219485. May 17, 2007. Accessed on August 25, 2010.

Schelling, Christianne. "Cat Scratching Solutions." **www .catscratching.com**. Accessed on June 25, 2010.

Schwartz, Marion. *A History of Dogs in the Early Americas.* Yale University Press: New Haven, CT. 2007.

"Sheep." Irvine Mesa Charros 4-H Club. **www.goats4h .com/Sheep.html**. Accessed on October 29, 2010.

"Snake Mythology." University of Massachusetts Amherst's Natural Resources and Environmental Conservation. **www.umass.edu/nrec/snake_pit/pages/myth.html**. Accessed on August 25, 2010.

Springer, Ilene. "The Cat In Ancient Egypt." Tour Egypt. **www.touregypt.net/egypt-info/magazine-mag 04012001-magf1.htm**. Accessed on June 25, 2010.

"The HSUS Offers Pet Safety Tips for Winter Weather." The Humane Society of the United States. **www.humane society.org/news/press_releases/2009/12/winter _pet_safety_120709.html**. December 7, 2009. Accessed on June 25, 2010.

Thrift, Anastasia. "All About Pet Bird Perches." BirdChannel.com. **www.birdchannel.com/bird-housing/ bird-housing-accessories/choosing-a-perch.aspx**. Accessed on June 25, 2010.

"U.S. Pet Ownership — 2007." American Veterinary Medical Association. **www.avma.org/reference/market stats/ownership.asp**. Accessed on October 1, 2010.

Walton, Marsha. "15,000 Years With Man's Best Friend." CNN. **http://archives.cnn.com/2002/TECH/ science/11/21/coolsc.dogorigin/index.html**. Accessed on June 25, 2010.

Wells, Virginia. "Why Do Cats Like Small Places?" Petplace.com. **www.petplace.com/cats/why-do-cats -like-small-places/page1.aspx**. Accessed on June 25, 2010.

"Wild Wolves." NOVA Online. **www.pbs.org/wgbh/nova/ wolves/05false.html**. Accessed on June 25, 2010.

"Winter Shelter for Pastured Horses." Equisearch. **www .equisearch.com/horses care/health/winter/eqshelter 3279**. Accessed on October 8, 2010.

Zollinger, Sue Anne. "Why Dogs Turn In Circles Before Lying Down." Moment of Science. **http://indianapublic media.org/amomentofscience/turning-circles-lying**. January 14, 2010. Accessed on June 25, 2010.

Index